# Atlantic Coast Beaches

## A Guide to Ripples, Dunes, and Other Natural Features of the Seashore

WILLIAM J. NEAL • ORRIN H. PILKEY • JOSEPH T. KELLEY

2007
MOUNTAIN PRESS PUBLISHING COMPANY
Missoula, Montana

Cover photo of sand dune courtesy of Drew Wilson/*The Virginian-Pilot*

**Library of Congress Cataloging-in-Publications Data**

Neal, William J.

Atlantic coast beaches : a guide to ripples, dunes, and other natural features of the seashore / William J. Neal, Orrin H. Pilkey, Joseph T. Kelley.

p. cm.

Includes bibliographical references and index.

ISBN-13: 978-0-87842-534-1 (pbk. : alk. paper)

ISBN-10: 0-87842-534-9 (pbk.)

1. Beaches—Atlantic Coast (U.S.) I. Pilkey, Orrin H., 1934– II. Kelley, Joseph T. III. Title.

GB459.4.N43 2007

551.45'7097—dc22

2007008919

PRINTED IN THE UNITED STATES OF AMERICA
BY F. C. PRINTING, SALT LAKE CITY, UTAH



Mountain Press Publishing Company
P.O. Box 2399
Missoula, MT 59806
406-728-1900
www.mountain-press.com

*To Mary, Sharlene, and Alice*

# Contents

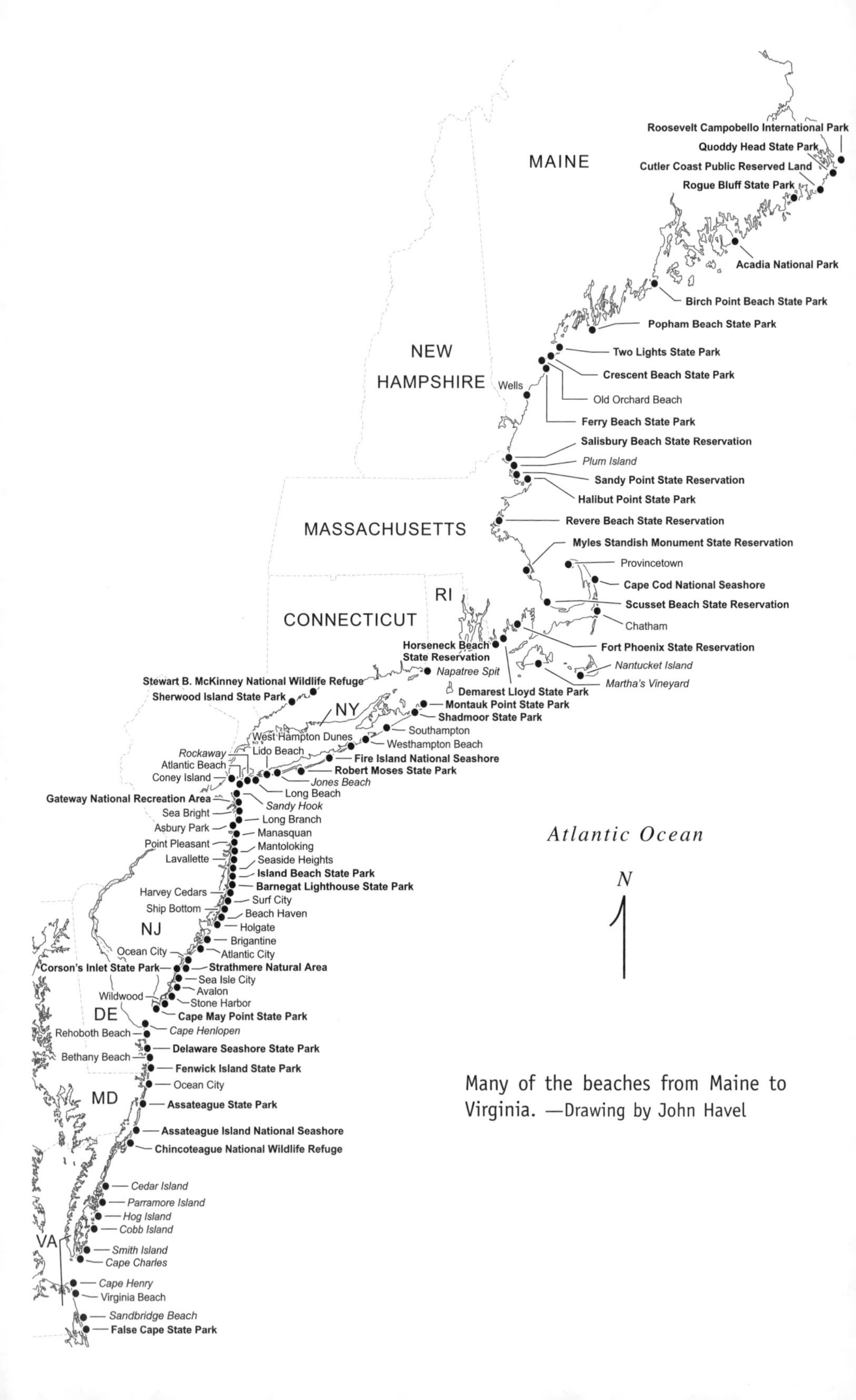


Many of the beaches from Maine to Virginia. —Drawing by John Havel

# Preface
## *A Seashell Bias*

Everybody knows about seashells. They're colorful, pretty, easy to collect, and they tell a story about the environment and its former inhabitants. The sand beach, where the shells reside, also has stories to tell. The problem is that most people have a seashell bias. Everyone knows about seashells, and almost everyone who has visited a beach has a special curiosity about shells. But while walking, head down, the shell seekers' eyes have moved right over a myriad of features that are just as interesting, features that reveal all sorts of things about the history of the beach and the processes that act on it. For example, erosional and depositional features of a beach, such as berms and runnels, barking sand, swash marks, and ripples, all tell us about past events on the beach. One purpose of this book is to erase some of this bias and educate readers about the multitude of beach features a visitor can recognize and enjoy.

Beaches are the buffers between land and sea, and therefore are dynamic places. They are amazing features capable of withstanding the biggest of waves and the mightiest of storms. But beaches don't just sit still and take what nature dishes out. They move about and change their shape in response to changing conditions. In fact, beaches are alive and seem almost capable of thinking and planning. When big storm waves arrive, the beach flattens itself by moving sand to and fro. The flattening causes the energy of waves to dissipate over a much broader surface area, which reduces the distance that the shoreline will move back, or erode, during the storm. After the winds die down and the waves diminish, currents and waves move sand back to the beach, the erosion is repaired, and the beach gets steeper again.

Beaches respond to processes of varying scale, which range from winter storms that may affect hundreds to thousands of miles of shoreline; to hurricanes that affect tens to hundreds of miles; down to the small events and processes that you see on a beach stroll, such as individual waves or small features produced by wave swash. In fact, a vast array of events are going on every day, every minute, and every second on the beach. Currents and waves are moving sand back and forth, shaping and etching bed forms, like ripple marks; wind is moving sand and piling it up in rows of dunes or behind obstructions; tides are moving air in and out of the sand as if the beach were a giant bellows; and animals are burrowing as they scavenge for food while birds prance across the surface trying to catch them among the sand grains. All of these events leave some sort of record in the sand.

Geologists long ago recognized that studying these records of modern beaches and the continental shelf—the gently sloping surface offshore and adjacent to beaches—could help us understand the origins of some ancient sandstones. "The present is the key to the past" goes an old geologic maxim, and so the study of modern coastal environments became an important aspect of the field of sedimentary geology.

Nowadays, in introductory geology classes, a trip to the beach (if there is one nearby) is a given. What geology students learn on beaches, however, seems to be relatively unknown by the general public, who seem to be more interested in pretty shells than beach-forming processes. At least that's the impression we have gotten from talking to beach strollers on many beaches over many decades. Yet we also have the impression that when the various features on the surface of beaches are explained to people, they are usually fascinated by them and want to know more. So that's why we have written this book, a layperson's guide to the beach that we would like to subtitle *Everything You Need to Know about Beaches Except Seashells!* (Truth be told, we do cover some important and widely overlooked aspects of seashells, such as their age, color and staining, orientation, and rounding from surf abrasion, because these things tell us a lot about how beaches work.)

Many beach features are poorly understood. We're not at all sure how some features form. For example, we thought that all of the small "nail holes" we saw in the upper beach formed when air passed through the beach. Then one day we saw a little feeler sticking out of a hole and realized that some holes are made by animals. As we point out in several chapters, interested tourists, beach dwellers, and students in need of scientific term-paper projects could learn a lot about how beaches work just by standing in the swash zone and watching nature at work. One could watch air holes or rings form in the beach and explain why they form in one place but not another. By walking the beach on successive days, one might begin to grasp why some sand squeaks or barks but other sand doesn't. Burying seashells in the mud could demonstrate how shells become blackened. Looking at the details of

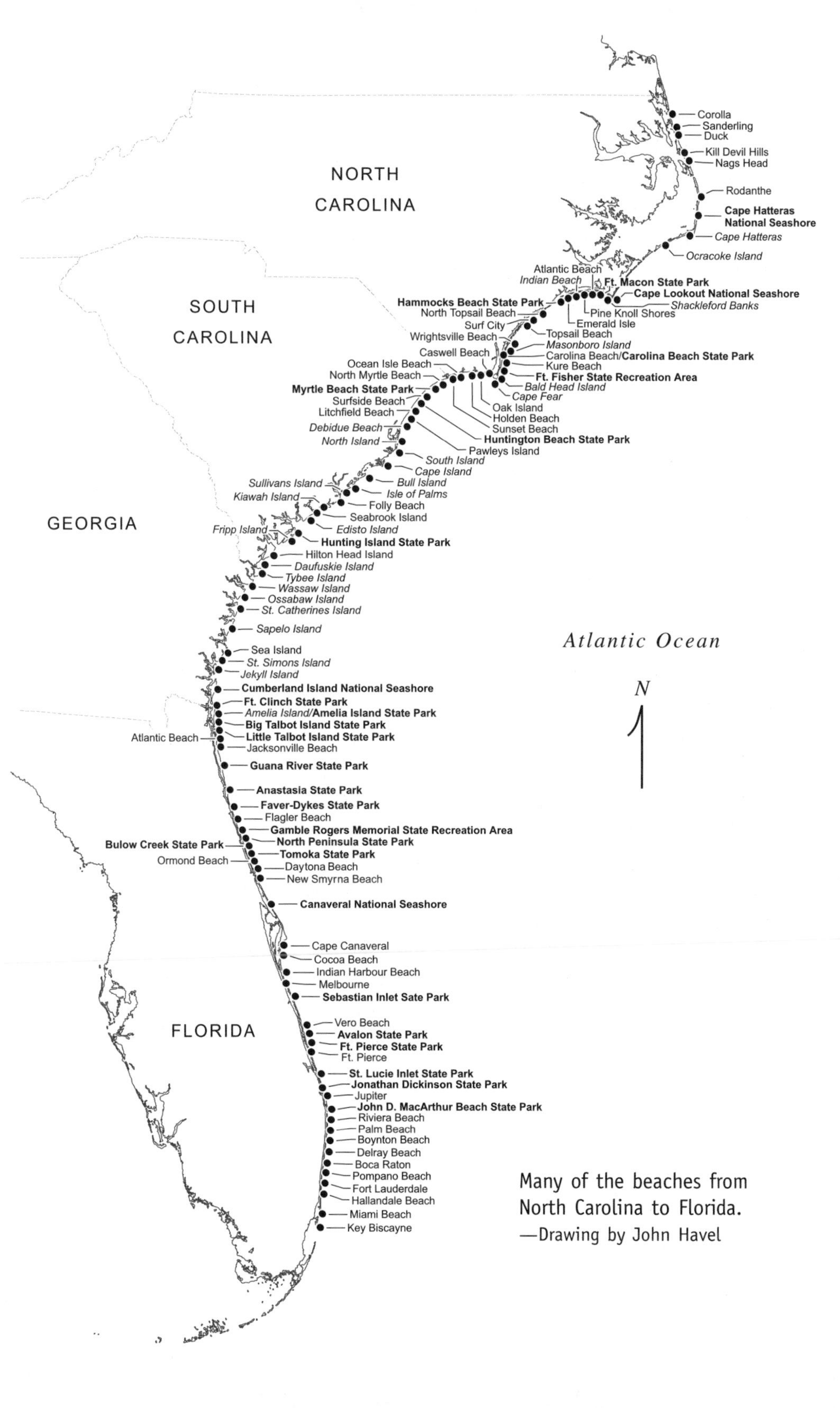


Many of the beaches from North Carolina to Florida.
—Drawing by John Havel

a beach before a storm and then afterward might open up a world of understanding of beach dynamics.

All three of us have written books on the hazards of living next to a beach. In these books we discussed the problems with engineered structures, such as seawalls, and the need to maintain the health and recreational quality of beaches for future generations. This volume is different. Although we do mention some of the environmental problems that East Coast beaches experience, our emphasis here is on natural processes. We hope you find this book fun and that you enjoy using it as much as we enjoyed compiling it.

We express our thanks to all those who helped with this endeavor. In particular, Drew Wilson, Sidney Maddock, Andy Coburn, Tracy Rice, and David Godfrey furnished some excellent beach photos and generously gave us permission to use them. Over the past year or two we have gleaned all kinds of information about local beach features along the East Coast from various colleagues. We are grateful for input from Andrew Cooper, Karen Dugan, Stewart Farrell, Charlie Finkl, Duncan FitzGerald, Jim Henry, Duncan Heron, Gered Lennon, and Bob Oldale, among others. Eddie Jarvis was our able field assistant and carried out an independent study on the size and frequency of sand holes on beaches from Florida to New York. Katherine Ort was our first contact at Mountain Press, and she helped organize the book, all the while referring to us as the "beach boys." Our editor, James Lainsbury, provided technical guidance, and we express our gratitude for his patience. Charles Pilkey drew most of the technical drawings of beach processes, and John Havel produced the index maps. We thank our reviewers, particularly Walter Barnhardt, Dave Bush, and Art Trembanis for their constructive suggestions.

We are grateful to a large number of geologists who have studied the bed forms and processes of beaches, dunes, and shallow nearshore waters. Much of this work was done prior to the 1980s. Although there are several atlases of sedimentary structures, we found the following to be the most useful references for our purposes: the two-volume series by J. R. L. Allen, *Sedimentary Structures: Their Character and Physical Basis*, and the textbook by H. E. Reineck and I. B. Singh, *Depositional Sedimentary Environments*. Both of these monumental efforts summarized much of the published work of their predecessors.

WILLIAM J. NEAL
ORRIN H. PILKEY
JOSEPH T. KELLEY

# Geology of the Atlantic Coast

The two basic features that distinguish one environment from another are the geologic and climatic setting. Beaches are no exception, so understanding a beach begins with an exploration of such settings. To most of us, the climatic aspect is obvious; daily weather is a focal point of our lives. Beaches are highly influenced by day-to-day weather and longer-term climate, including wind-driven waves, the daily rise and fall of the tide, hurricane storm-surge flooding, and the currents formed by wind direction and water masses of differing temperatures. The geologic setting may be less familiar, and its influence on the type and character of beaches may not be as obvious.

To a degree, all features of a beach—from the micro to the regional scale—are influenced by the geologic setting. Our continent is a complex of many rock types, which have formed over a 3.5-billion-year history of multiple episodes of deformation, continental uplift, the wearing down of mountain ranges, and the transfer of sediments from the continental highs to the topographic lows of the coastal margins. The resulting contrasts between such geologically distinct provinces as the glaciated New England Highlands, the unglaciated Southern Appalachians, the Piedmont, and the relatively flat Atlantic Coastal Plain account for different kinds of beaches in terms of size, shape, distribution, and right down to the mineral composition of individual sand grains that make up a beach.

You will see that beaches are like people. They each have unique personalities shaped by beach material, the surrounding geologic and oceanographic frameworks, and climate.

## Different Kinds of Beaches

Beaches are the geological landforms that many people are most attracted to. Beaches are the focal point of vacations, and the travel media tout the nation's "best beaches." The number of beach vacationers swells annually. And more

Aerial photo of a Georgia barrier island beach at low tide. Barrier islands are separated by channels called *tidal inlets,* backed by a salt marsh, and often have an inland maritime forest where not disturbed by human development. The long spit indicates that the longshore current travels toward the lower right corner of the photo.

and more people want year-round access to the beach, so the distance to beaches is among the criteria some of us use to choose where to live. In 2003 the population of U.S. coastal counties stood at 153 million, or 53 percent of the total population. Much of that population is concentrated along the Atlantic Seaboard, including some of the fastest-growing counties in the United States. Atlantic beaches have a magnetic draw. Few of these visitors, however, understand the dynamics of beaches, their origins and evolution, their fragility, or the details of daily changes that take place underfoot.

The story of U.S. Atlantic beaches begins with their geologic setting. Beaches can look very different from New England to New York to New Jersey and on to Florida, yet they are all beaches! From Long Island (New York) to Florida, the majority of beaches are the shorelines of barrier islands, which we refer to as *barrier island beaches*. These beaches are usually backed by dunes or overwash sand flats.

Cape Cod's sandy strands offer recreational opportunities similar to those of barrier islands, but the overall appearance of these beaches differs. The easily accessed beaches at Cape Cod National Seashore, for example, are not on islands. They have water on their ocean sides, but their landward side is

The beach at Marconi Station in Cape Cod National Seashore (Massachusetts) is called a *mainland beach* because it is attached to the upland and is forming as the upland erodes. The upland here consists of glacial deposits, which are the source of much of the sand on the beach.

bounded by high bluffs of sand and gravel unlike anything seen on barrier islands. Known as mainland beaches, they formed in an altogether different way than barrier island beaches, and they possess many features not found on beaches to the south.

Spits are beaches that are similar to both mainland and barrier island beaches. Like mainland beaches, spits are connected to the land at one end, and in New England there is often an eroding bluff supplying sand to the beach. Like barrier islands, spits end in the water, usually at a tidal inlet, for example, Sandy Neck (Massachusetts). When a spit is cut in half by a storm, a barrier island is formed.

Tombolos are beaches that connect the mainland with an offshore island. They are more common in New England, with its numerous offshore islands, than along the mid-Atlantic or southeastern U.S. shores.

Elsewhere in New England, beaches are commonly found in association with outcrops of bedrock. Called *pocket beaches*, they are named for the fashion in which they nestle between rocky headlands, as a hand fits into a pocket. Just as they differ in appearance from barrier islands and mainland beaches, pocket beaches differ in origin and in some of their features. For

Sandy Neck (Massachusetts) is an example of a large spit. Sand is carried from the mainland beach (far background) and deposited in the inlet at land's end. The feature continues to grow by sediment accumulation, but tidal currents cause the end of the spit to curve into the embayment to the left. In the right foreground a smaller spit is growing off the bend of the larger spit.

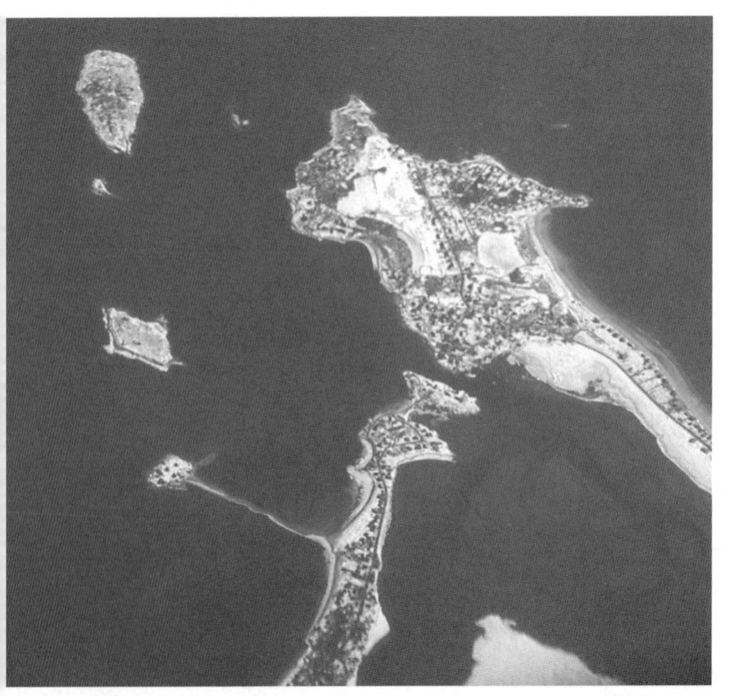

There are several tombolos, or beaches that connect islands to the mainland, in this air photo. The largest tombolo extends from the mainland, right of this photo, to the large island and the community of Biddeford Pool, Maine, at the top of the photo. Another tombolo comes from the bottom and connects several small islands near the center of the image.

example, they often consist of gravel or cobbles rather than sand. Nevertheless, all of these settings are beaches.

Many beaches have become so popular that visitors just don't want to leave. Developed beaches may have once been like any of the other three types of beaches, but today they are dominated by houses, boardwalks, hotels, condos, and often seawalls. Many of these developed beaches are now artificial, meaning they are constructed of sand dredged or pumped from offshore or sand trucked from inland sources. Despite their changed appearance, developed beaches experience the same forces and share the same history as natural beaches, at least until human structures interfere with normal beach and coastal processes.

Swimming, sunbathing, surfing, fishing, shelling and beachcombing, or just plain lazing about are activities everyone enjoys at the seashore. Discovering similarities and differences among the beaches we visit, or the significance of the features we see on our favorite beach, can add another element of enjoyment to our visit. Helping to answer the question "Why do beaches differ in size, shape, color, and many other regards?" is one of the goals of this book. Each beach had its own unique origin, has different orientations and kinds of sediments, as well as different volumes of sand that are supplied to or eroded from the beach. Different wave types and currents, different storm histories, and different tidal ranges all cause beaches to differ from each other. And just as the plants and animals of the northern forests

Sand Beach in Acadia National Park (Maine) is a typical New England pocket beach, so named because this type of beach forms in a pocket between rocky headlands and is not laterally extensive.

The beach at Virginia Beach is an example of a mainland beach backed by heavy development. As will be noted in later chapters, such beaches are prevented from migrating landward in response to sea level rise and must be maintained artificially.

differ from those of the south, so the creatures and plant communities (and the tracks and traces they leave) on New England's beaches differ from those in Florida. On a beach walk one can unravel the fascinating story that each beach has to tell.

### *The Geologic Setting*

The path to understanding beaches is as serpentine as the Atlantic's shoreline. Regional divisions in shoreline character are dictated by a long geologic history. The geological divide of the Eastern Seaboard runs in an east-west direction through Long Island (New York). North of Long Island, New England's shoreline is generally rocky and highly irregular in outline. Bedrock of the Appalachian Mountains extends down to the shore and out beneath the sea of this region. In Acadia National Park (Maine), for example, granitic Cadillac Mountain, 1,500 feet in elevation, extends into water that is hundreds of feet deep directly offshore. These high-relief, erosion-resistant rocks really create the unique character of the New England coast. They form numerous smaller islands and shoals and rocky headlands that profoundly influence sediment transport and the accumulation of beach sand.

From Long Island's South Shore all the way to South Florida, the Atlantic Coast is more linear and regular; it's mostly fronted by chains of barrier islands. Behind the islands the mainland consists of sediments that were eroded from the Appalachian Mountains and deposited as an apron of sand and mud at their base. These island chains are occasionally broken by large estuaries, like Delaware and Chesapeake Bays; smaller estuaries, like those behind the barrier islands in Georgia and South Carolina; and large bodies of water behind islands, such as North Carolina's Albemarle and Pamlico Sounds. Along most of Florida, the barrier islands are separated from the mainland shore by narrow lagoons.

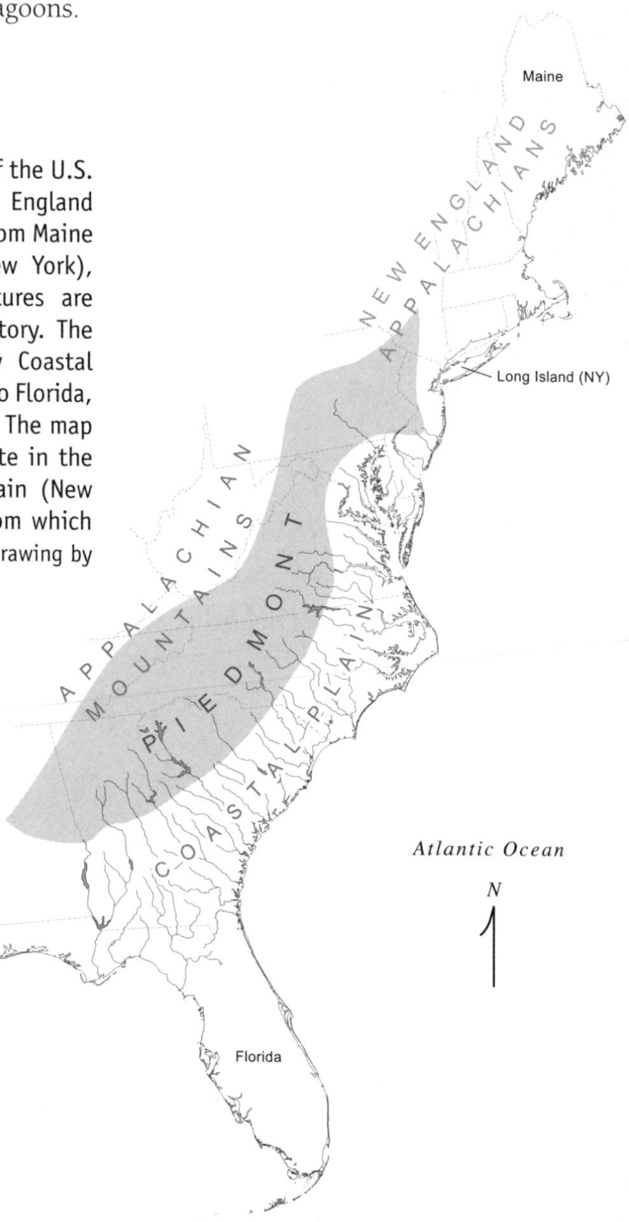

The physiographic provinces of the U.S. Atlantic Seaboard. The New England Appalachians form the coast from Maine to north of Long Island (New York), a region where coastal features are directly related to glacial history. The shoreline of the sedimentary Coastal Plain, from Long Island south to Florida, is fringed with barrier islands. The map also shows rivers that originate in the Piedmont and the Coastal Plain (New England rivers not shown), from which beach sands were derived. —Drawing by John Havel

The offshore ocean bottom also differs from north to south. The northern continental shelf, the Gulf of Maine, is almost 1,000 feet deep in some of its irregularly distributed basins but contains ledges and banks that are less than 200 feet deep. Along the Gulf of Maine's outer margin is Georges Bank, a shallow platform that drops off abruptly at the edge of the continental shelf into the Atlantic abyss. In contrast, south of New England the continental shelf is a relatively smooth ramp that dips gently to its termination (at around 300 feet below sea level) at the edge of the continental slope. The continental slope descends off into the deep sea to the flat abyssal plains that are 3 miles below the sea's surface. This gentle continental shelf played a role in the origin and distribution of the Atlantic's barrier islands and their associated beaches. The width and relatively shallow depth of the continental shelf is a major control of the size of waves on East Coast beaches. A narrower shelf means deeper water near the coast, allowing higher waves to strike. Where the shelf is wide waves expend energy moving shelf sand, so waves are smaller when they reach the shore.

The Long Island border between these northern and southern provinces contains a bit of both regions' characteristics. Though lacking bedrock, Long Island's northern side is irregular with numerous small coves and sand and gravel spits. Its southern side is more like shorelines to the south in that a barrier island chain borders the coast for more than 100 miles. Off the Hudson River there are relatively deep valleys cut into the sandy material of the continental shelf. The edge of the shelf is incised by submarine canyons that date back into the not-too-distant geologic past when sea level was lower and the shoreline was near the outer edge of the present continental shelf.

### *The North-South Dichotomy*

What has led to such a dichotomy in the shoreline of eastern North America? The entire coast, from northern Labrador to Florida, shares a broadly similar, complex geologic history.

The theory of plate tectonics is a fundamental part of modern geology. It states that the outer layer of the earth, of which the continents are a part, is broken into separate plates. These plates move in different directions on the earth's surface due to the transfer of heat from the earth's inner recesses. Plates collided in the past, forming supercontinents that eventually broke up into smaller continents. When plates collided, mountain ranges formed. Today's continents are the result of these processes.

As early as 320 million years ago, the continental masses of Europe and North Africa were colliding with North America and forcing the Appalachian Mountains to rise. At the time of their formation, this range of peaks may have towered as high as the modern Himalayas. Beginning about 225 million years ago, the supercontinent Pangea, which had formed in that collision and included Europe, Africa, North America, and South America, began to split apart. The rift began in the north and proceeded south as the modern Atlantic Ocean was born. We are passengers on the North American plate, still drifting apart from Europe at about ½ to 1 inch per year.

The Mid-Atlantic Ridge is the rift boundary between these two great crustal slabs and comprises a chain of volcanic mountains on the seafloor that continue to evolve wherever lava upwells and pushes the two plates away from each other. The eastern side of the North American continent is referred to as a *trailing edge* because it is tucked in behind a continental mass that is drifting away from a rift zone. This pattern of movement, in part, determines the types of coast that characterize the Atlantic Seaboard. Trailing-edge coasts have coastal plains bordering the sea. Such gently sloping coasts are often characterized by barrier island beaches or long mainland beaches without rocky cliffs or headlands. In contrast, the North American Pacific Coast is a leading-edge coast characterized by a rugged shoreline.

Because of their great height, the early Appalachian Mountains eroded quickly, shedding copious quantities of gravel, sand, and mud into rivers that brought the sediment to the newly forming shoreline of the ancestral Atlantic. As the Appalachians eroded, the trailing margin of North America built seaward into the ever-widening and ever-deepening new ocean. Over tens of millions of years, sea level rose and fell, owing to geological changes in the shape of the Atlantic Ocean basin as the plates moved about. The position of the shoreline shifted landward or seaward in response to these sea level changes. When volcanoes were most active along the Mid-Atlantic Ridge, the undersea mountains displaced vast quantities of seawater as they bulged up and out, and world sea level rose and flooded the continental margins. When the mid-ocean volcanoes were less active, they subsided; world sea level fell and rivers built the continental margin seaward again.

The great mass of sediment that was deposited on the margin of North America by Appalachian rivers and spread back and forth by migrating shorelines is today's Coastal Plain. The Coastal Plain continues below the present level of the sea, where its surface is called the continental shelf. All of this area has been alternately flooded and emergent through millions of years of earth history. So much material has been added to North America that the Appalachian Mountains are now miles to hundreds of miles back from the ocean, and sediments, more than 15,000 feet in thickness, compose the Coastal Plain and continental shelf.

Although the entire eastern edge of North America shares this geologic history, the Coastal Plain generally is missing north of New York. There are scattered Coastal Plain remnants; for example, a 40-million-year-old peat deposit was quarried for many years in Vermont. People occasionally dredge up 30-million-year-old plant fossils while fishing over ledges in the Gulf of Maine. By and large, however, the ancient rocks of the old Appalachian Mountains, hundreds of millions of years older than those of the Coastal Plain, make up the coastal region north of New York. The structure of the ancient mountains shapes that area, producing the distinctly rocky and rugged topography that is absent on the Coastal Plain.

### *Continental Glaciation*

The main reason the Coastal Plain is missing in the north has to do with ice. In the late nineteenth century, geologists began to systematically take note of the Atlantic Coast's regional differences and found one striking distinction between the north and south. North of Long Island, bluffs of boulders and clay, now recognized as the deposits of glaciers, abound; none are found to

An island near Kennebunk, Maine, was composed of this till, which was deposited under a mile of ice. Note the material of differing sizes, including sand, cobbles, and mud. By 2006 the island was gone, having been eroded to produce sand and gravel for nearby beaches.

the south. Much of Long Island is, in fact, a long pile of material deposited by glaciers; it is a moraine system with associated glacial outwash that marks the southernmost extent of the last glacial episode, which occurred around 22,000 years ago. Where glacial debris forms the northern side of Long Island, the type of sediment, which varies in grain size and composition from place to place, controls the shape of coves and headlands. These deposits also provide the sand and gravel that waves transport along the shore to form spits. The

---

**COMMON GLACIAL DEPOSITS**

TILL — A complex mixture of unsorted mud, sand, and gravel that was deposited directly by a glacier. Gravel-sized material ranges from pebbles to boulders. Tills with high mud (clay) content tend to be impermeable. Tills are found in moraines and drumlins, two common New England landforms.

MORAINE — A glacial deposit composed of till that occasionally has layered sands and gravels. Moraines may be deposited under a glacier, called a *ground moraine*, or at the margins of a glacier, for example, an end moraine. Marginal moraines occur as elongate hills and may be tens of miles long and have up to 100 feet of topographic relief, typically with hummocky topography. Impermeable till often results in wetlands between the hills.

DRUMLIN — An elongate hill, usually hundreds of yards in length and teardrop-shaped in map view, that was deposited by a glacier and then overridden by it to produce a streamlined outline. A drumlin has a steep end in the direction from which the ice flowed and tapers in a gentle slope in the direction the glacier flowed. Drumlins tend to occur in clusters, or drumlin fields, and most are composed of till; some were formed over a core of bedrock. Their exact origin is still unclear.

OUTWASH — Sediment deposited by glacial meltwater and rivers draining melting ice. These sands and gravels are layered (stratified), porous, and permeable. The most common New England landforms composed of this type of sediment are outwash plains, which are gently sloping, coalescing fans of sediment that extended away from the margin of ice.

GLACIAL-MARINE SEDIMENT — A general term for glacially derived sediment that was deposited below sea level, including muds that were deposited by meltwater that issued from tunnels in glaciers that were below sea level. Such mud deposits blanket landscapes and form gentle topography.

---

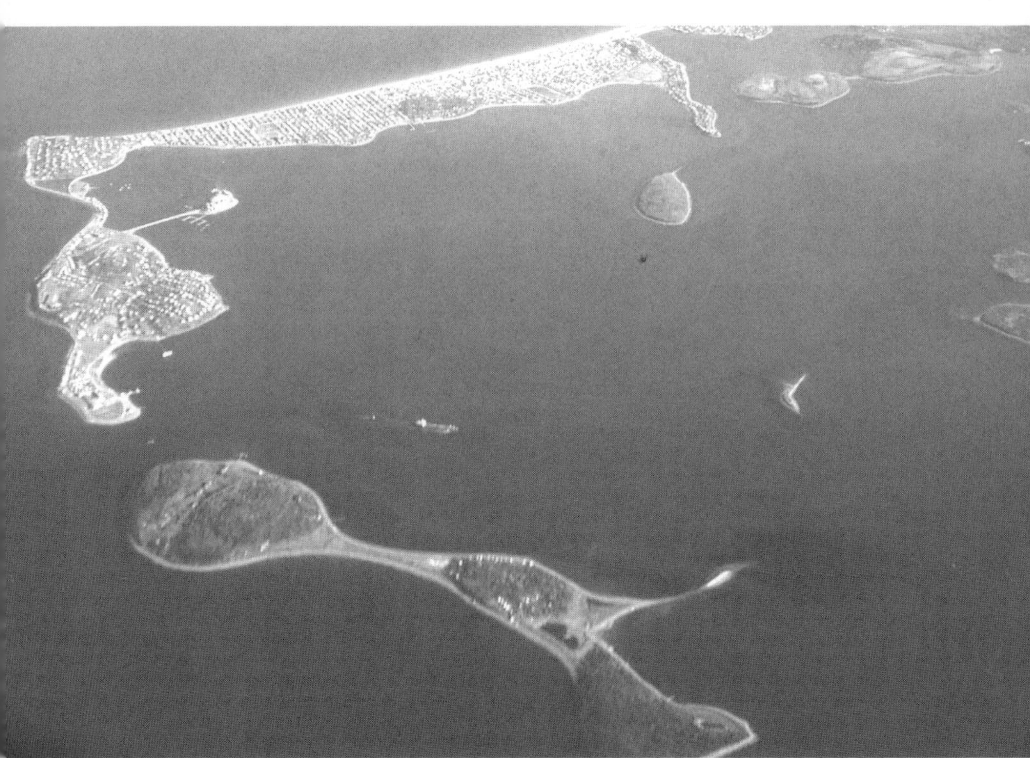

Aerial view of drumlin islands in Boston Harbor. Peddocks Island is in the foreground, and Hull, Massachusetts, is on the island to the left. The beaches of these islands are composed of material eroded from these drumlins and other drumlins that are completely gone.

southern and seaward side of Long Island is more uniform, created by sand from rivers that flowed out from the melting ice into the sea.

The glaciers obtained the material to form these deposits by eroding the land beneath them. Geologists now realize that there were many glacial episodes, perhaps one every 100,000 years or so, between 2 million and 10,000 years ago, a period of time known as the Pleistocene Epoch, or the Ice Age. On land there is a good record of only the last advance and retreat of the ice because each glacial episode eroded most traces of earlier events on land. In the deep sea, however, there was no erosion, and the sedimentary record of multiple glaciations persisted. It is clear that glaciation stripped the Coastal Plain sediments from the northern region, depositing them south of the limit of glaciation. The isolated pockets of Coastal Plain sediments that survive are small remnants; only at the edge of what was the ice's southern terminus, on Georges Bank (Gulf of Maine), do we find a northern Coastal Plain remnant still intact.

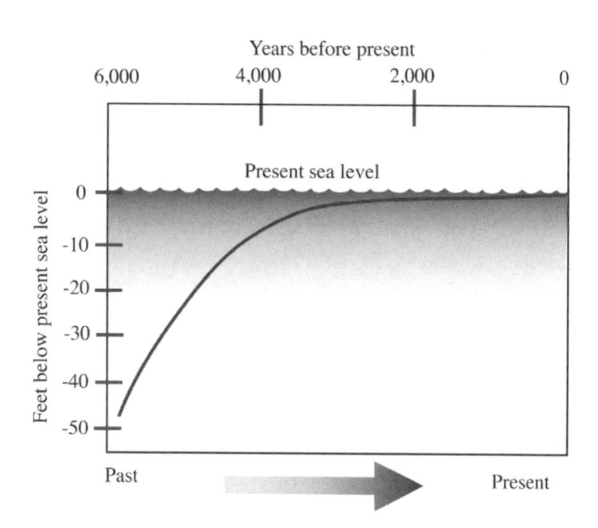


General sea level curve for North America over the last 6,000 years. Sea level began to rise rapidly over 15,000 years ago, and that rate continued until a few thousand years ago when it began to decrease (flattening of curve). Depending on their distance from the ice, many locales experienced dramatically different sea level curves than what is represented here. —Drawing by Charles Pilkey

Although the glaciers left New England and New York by 11,000 years ago and today we are living in an interglacial time, their influence lingers. Not only Long Island but all of Block Island (Rhode Island) and Cape Cod (Massachusetts) are giant deposits of glacial sediment. The glacial legacy is also distributed throughout New England in a myriad of glacial depositional and erosional landforms and sediments. One artifact of the Ice Age, however, continues to strongly influence both northern and southern coasts. Sea level and land level changes owing to the huge volume of water exchanged between the ocean and the ice, and the weight the ice imposed on the northern crust, continue to affect shoreline position. Over the long haul of geologic time, an ice age is likely to return; but for the immediate future of humans and beaches, the story is that of global warming. Whether natural or anthropogenic, global warming is driving a continuing rise in sea level.

The effect of sea level rise that began at the end of the last glacial episode is more striking where the Coastal Plain remains. Considering the very gradual surface slope of the Coastal Plain, a 1-foot rise in sea level can cause the shoreline to shift laterally by as much as 2,000 feet. Given that melting ice added enough water to the ocean to raise its level more than 300 feet in 20,000 years, one can see why the continental shelf is submerged and why geologists regard the present position of most beaches as temporary. Indeed, sea level in the more distant past, such as the warm period that preceded the last glacial episode, was higher than today. Old shorelines existed far inland from the present coast. The Surry Scarp, for example, is a series of fossil beaches that are traceable from Delaware to Georgia.

In addition to the significance of geologic setting on the character and evolution of beaches, other important influences on beaches include climate, water depth, physics, chemistry, and the biotic life of oceans. The U.S. Atlantic Seaboard spans just over twenty degrees of latitude, ranging from temperate to subtropical conditions that determine the types and abundance of plants and animals along the coast. These in turn control such beach aspects as the types of dunes that form and the amounts of carbonate sand (seashells) in beaches. The interaction of climate and oceans also accounts for regionally different wave energies, circulation patterns, tidal ranges, and storm impact on beaches. We discuss these things in more detail throughout this book.

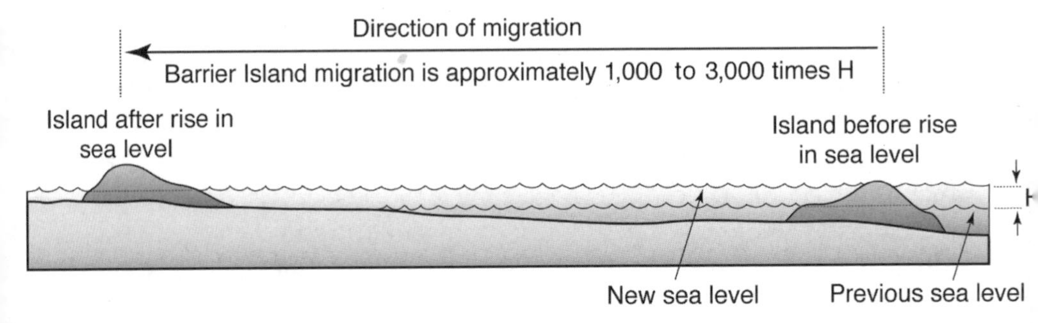


Depending on the upland slope, a 1-foot rise in sea level (H) can cause the shoreline position to shift landward 1,000 to 3,000 feet, as happened across the nearly flat Coastal Plain. —Drawing by Charles Pilkey

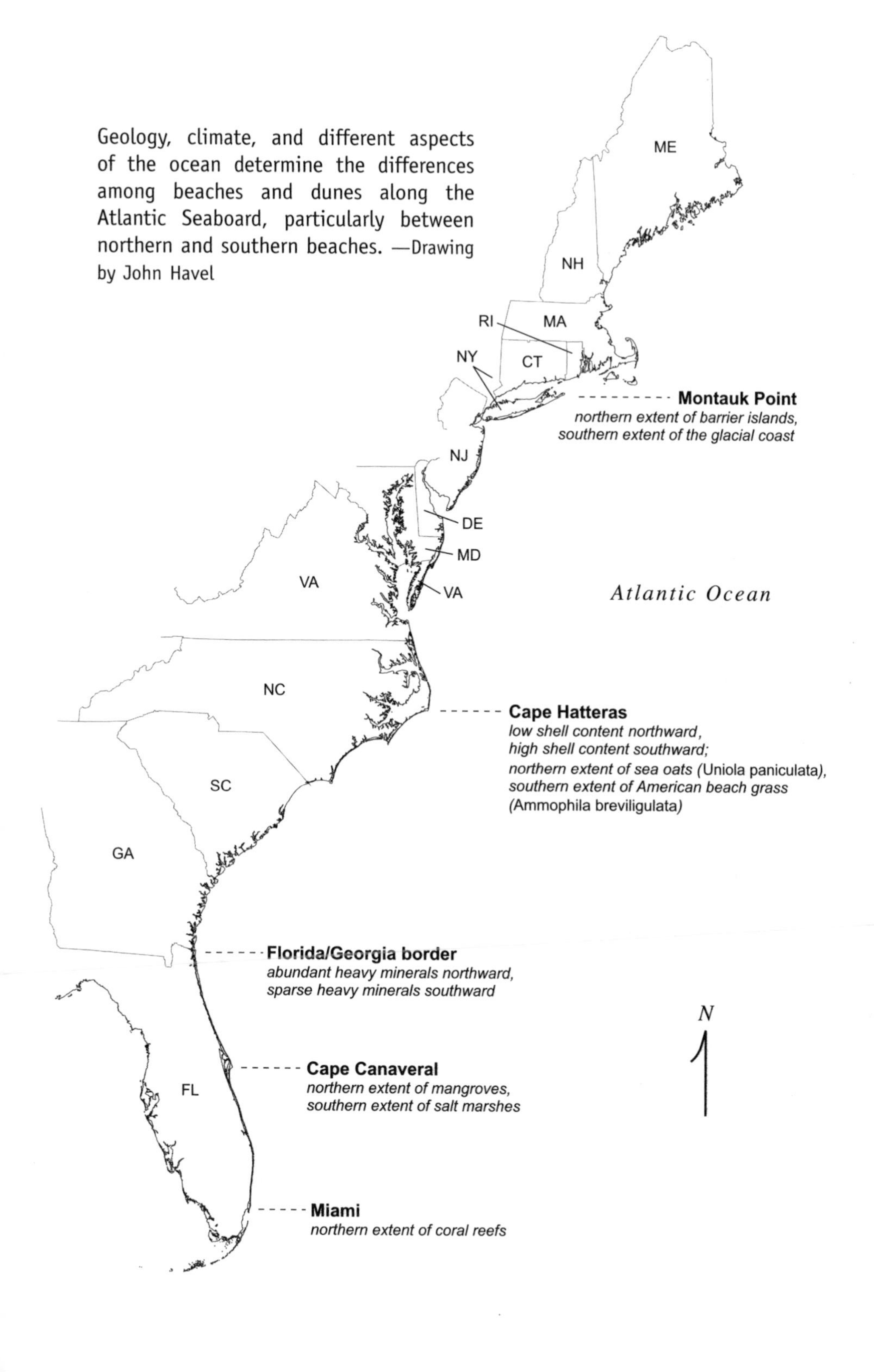


Geology, climate, and different aspects of the ocean determine the differences among beaches and dunes along the Atlantic Seaboard, particularly between northern and southern beaches. —Drawing by John Havel

# Beach Movement

We pose a riddle: What is changing constantly but appears to be timeless? The answer, of course, is the beach. Beaches represent one of the most dynamic environments on Earth. A dynamic environment is one in which high levels of energy are expended over short periods of time, causing rapid change. For beaches, energy comes primarily from waves, wind, tides, and currents.

The greatest change in a beach is its response to the rise in sea level. As sea level rises the beach must shift upward and landward, or be inundated. The latter is what happens to rocky shorelines like those of New England, which passively drown, although erosion may continually provide new material for continued beach formation. The rocks, however, have no ability to change their position or shape to accommodate the rising water. With time, the shoreline and beach position, the boundary between land and sea, will shift landward, or *retreat*.

In contrast, along the Coastal Plain, low-lying barrier islands are backed by salt marshes and lagoons, and storm waves carry beach sand across the island onto the marsh or into the lagoon. In the same way, the wind may carry beach sand to the back of the beach to form dunes that can continue to march to the back side of the barrier island. Along the Coastal Plain, the beach, dunes, and the entire barrier island gradually *migrate* landward as distinct landforms. Of course, if you look only at the shoreline position, you can say it has retreated landward as well. In both retreat and migration, the beach moves landward, and any static object such as a hut, house, or hot-dog stand in the path will be destroyed by the erosion that is part of the process. Strictly speaking, *beach erosion* ought to be used to describe the actual removal of sand from the beach that produces erosional features, such as scarps cut in the beach or the toe of a dune at the back of the beach; but the term is often used as a synonym for *beach retreat*.

The processes and responses along beaches are broadly predictable, but there are often surprises; for example, when storms strike. Beaches change very rapidly to adjust to different conditions, yet the beach is there from day-to-day, year-to-year, and over a lifetime, or so it seems. In actuality, many undeveloped beaches are moving in a landward direction at a rate of 2 to 5 feet per year, but as they move their appearance generally remains the same. If you own a cottage next to a beach, however, you are painfully aware of the rate of beach movement and call it *erosion*. Sometimes beaches build seaward for a while, but this is usually a temporary event. Without a marker such as a cottage to measure by, big changes in beaches may go unnoticed, but they are always moving.

The following sections examine different beach and coastal features in the context of time and space. We focus on the origin, evolution, and classification of barrier islands, because these islands dominate much of the U.S. Atlantic Seaboard south of the South Shore of Long Island. In response to sea level rise, barrier islands have migrated for miles across the continental shelf, so sea level change is something we review. As noted, this retreat is often referred to as *beach erosion*, even though beaches persist, shifting landward by different processes.

The rate of sea level rise is expected to increase in coming decades, so there are exciting times in store for beaches, especially those hemmed in between the sea and rows of shoreline buildings. On the other hand, beaches and barrier islands without buildings will be free to move landward as the sea rises, an amazing feat for such large natural features.

## Time and the Beach: The Short Life of Sand Castles

Travel brochures often speak of the "timelessness" of beautiful beaches. To geologists, however, nothing could be further from the truth! Natural beaches are dynamic, constantly changing; but at the same time they are depositing a record of that history of change. New England's rocky shores and the mid-Atlantic and southeastern U.S. barrier islands do reflect a long history of events that occurred over thousands of years, from the most recent glacial advance to the present, but that is not to imply that the beaches themselves are of such an old age. The actual beach form that you walk over may be hours to days old, perhaps the time span since the last storm, even though the beach may have been in place for centuries. Other features on the beach, like sand ripples, record events that are very recent and of short duration (minutes to days). The grains of sand, for example, quartz, which make up the beach may have formed as crystals in rocks hundreds of millions of

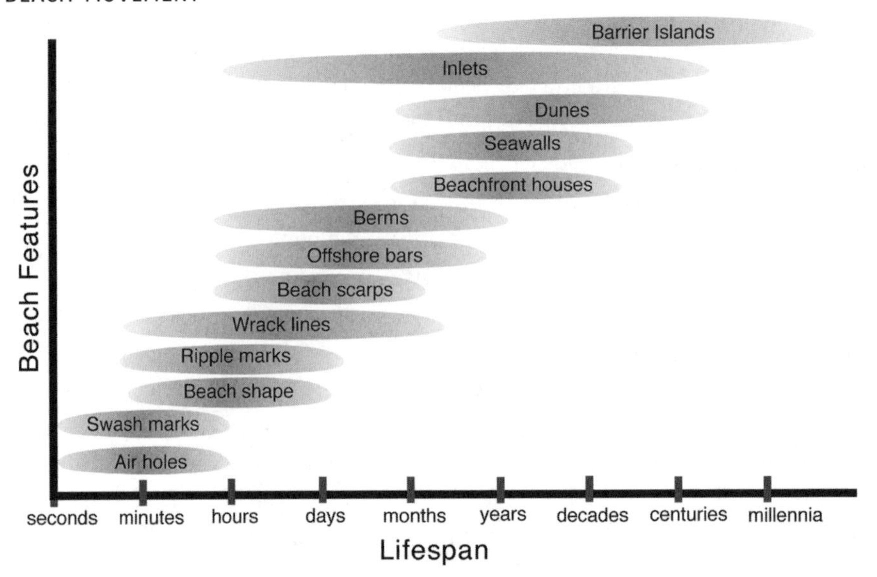


The typical life span of beach and coastal features ranges from seconds to millennia. Larger natural features, such as inlets, dune fields, and barrier islands, persist for much longer spans of time than the small features of the active beach, such as ripple marks, swash marks, and air holes. —Drawing by Charles Pilkey

years ago while the shell fragments in the sand may have formed last year. So beaches do contain a record of the events that shaped them in the past, both long-term and very short-term. The record is found in a myriad of structures and objects that beach visitors frequently walk past without noticing, many of which we discuss in this book.

To illustrate the dynamic nature of beach features and their life spans, consider the example of a sand castle. If children build a castle just above the water line at high tide, the castle will persist at least until the next high tide. Although the castle may be degraded as people walk over it, the castle's very presence indicates that someone was here playing in the recent past. If the same children build a castle near the low tide line, the life span of the structure will be very brief because waves will wash it away as soon as the tide rises.

In a similar way, nature creates features, large and small, that tell an observer something about the history of the beach. A feature that is very small, above ground, and between the high and low tide level, like a castle, lasts only a brief time. Later in the book we describe small structures produced by wave swash, marks left by air bubbling from the beach, and ripples in the sand produced by waves or currents, any of which usually persist for only seconds

This New England beach and spit at low tide show evidence of the continuum of change. The offshore bars will change shape with each tidal cycle (hours), and the position of the dark wrack line on the beach changes position with the highest tide (hours to days). Older features include overwash fans (lobe-shaped extensions of sand into land interior) from the last great storm or hurricane that inundated the island. These fans are now part of a grassy plain where a dune line is forming (years to decades) as the vegetation traps sand. The maritime forest is the oldest environment identifiable in the photo (centuries to millennia). The boxed area is the ground represented in the photo below.

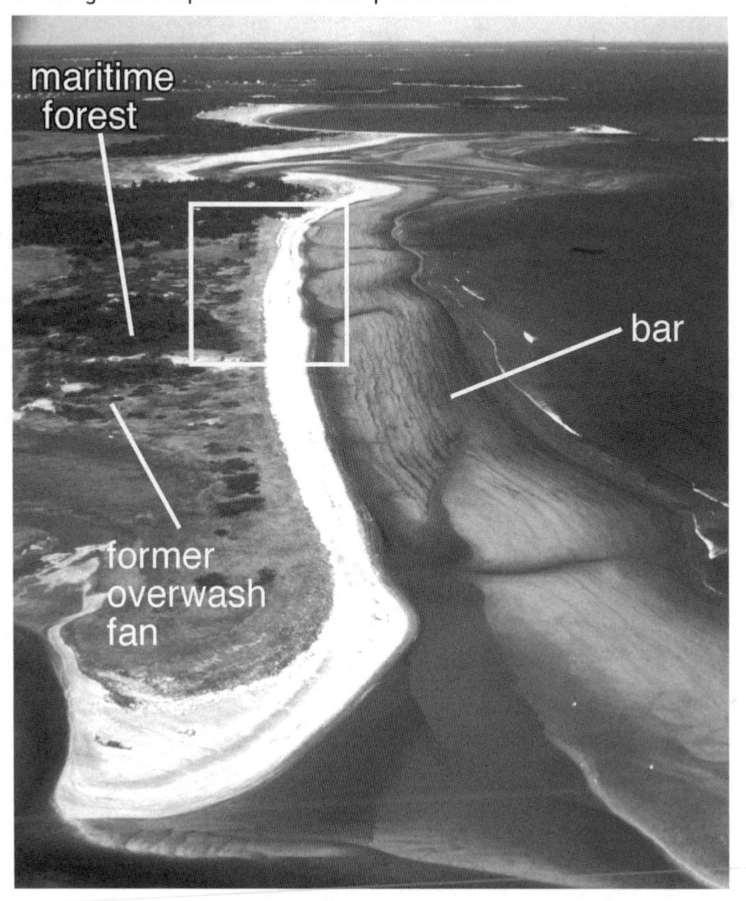


Ground view of the boxed area in photo left showing the position of a 1978 storm scarp in the dune. This 1992 photo shows the old scarp is gradually being buried by new dune growth. The beach is changing hourly, but beach sand is only carried into the dune-overwash zone by occasional big storms that occur years to decades apart.

The bedrock of Roque Bluffs on the coast of Maine is hundreds of millions of years old, while the sediments in the bluffs along the shore were deposited by glaciers around 14,000 years ago. The island, previously just an elevated area of land, became an island as the glaciers melted and sea level rose at the end of the Ice Age. Once an island formed, coastal processes began to form the bluffs, rocky shores, and beaches, which have persisted as features for a few thousand years but continue to change almost daily under the onslaught of the sea.

to hours. If preserved in the rock record, however, such features can persist for hundreds of millions of years. Although very few fragile beach structures survive to become features of sedimentary rocks, such a large number of structures are produced through time that some do get buried and preserved in the rock record.

Something that is much larger than a ripple and more removed from wave action, like a sand dune, provides a record of events that occurred for years, decades, or even centuries. It takes a long time for fields of sand dunes covering several square miles, and their associated grass and shrub communities, to develop. Because dunes are large enough to withstand small day-to-day wind and wave events, extensive areas covered by such dunes can be of considerable age. Behind the dunes there can be an extensive maritime forest, which may represent one of the oldest components of a barrier island system. A maritime forest has very old trees, and the original land on which the forest originated might be a few thousand years old. Natural barrier islands often exhibit all of these features.

Topographic features such as dune ridges and other shoreline features that were incorporated into the back-beach area of a mainland or barrier island beach may be obscured from today's visitor by a dense cover of maritime forest. In other words, beaches could have been present in what is now the interior of a forest.

As old as some beach features are, they are not really old in terms of the earth's history; beaches are not timeless. The bedrock of the islands along the Maine coast, for example, is hundreds of millions of years old, but the islands themselves have only existed for thousands of years. They were created as sea level rose and water surrounded the hills and other elevated areas of bedrock along the coast. A pocket beach on one of these island's shores may have been present for only a few decades or for hundreds of years, but the beach itself changes annually, daily, hourly, and even by the minute. We should not confuse the vast scale of the earth's history over geological time with the brief scale of features on a beach; nothing—be it condos, ripples, or dunes—lasts very long on the beach.

## Sea Level Rise: Time's Tide

Few aspects of our world seem more constant than the sea. Covering three-fourths of the earth's surface with depths averaging close to 3 miles, the ocean's size alone buffers it from small changes. The level of the sea has been regarded as stable throughout much of history, and people have used it for a long time as the datum from which to measure all elevations above (topography) and below (bathymetry) the sea's margin. Drowned Roman and Greek ruins along parts of the Mediterranean Sea, however, indicate that sea level is rising; yet in Scandinavia and other high-latitude locales, sea level has historically fallen. This apparent contradiction is primarily related to former ice cover. During the Ice Age, land areas covered with ice were depressed (pushed down) due to the ice's mass. When the ice melted the land surface rebounded and rose, resulting in a fall in sea level in these locales.

Such global events have affected the ocean in the past several decades, and changes are taking place in the level of the sea now. The 1964 Alaskan earthquake raised coastal lands in southeast Alaska, thereby lowering local sea level. At the same time, melting Alaskan glaciers have contributed to global sea level rise. Although modern rates of change in sea level are slow, the level of the ocean has changed enough in the recent past so that we no longer use sea level as the formal reference for all elevations and depths. The United States now uses a fixed reference point, the National Geodetic Vertical Datum, from which to reference such measurements.

Aerial photo of a raised delta in Columbia Falls, Maine. The sea once stood at the same elevation as the flat top of the delta, which today is a blueberry field. Snow marks the former shoreline along the outer edge of the delta. The melting glacier occupied what is now a pond to the upper left.

Melting glaciers have had the most influence on sea level change during the past 20,000 years. Since reaching their maximum southern extent 20,000 years ago, most of the continental glaciers of the northern hemisphere, including those of North America, have melted away, adding a huge volume of water to the ocean. The glaciers were more than a mile thick over much of northern North America, northern Europe, and northern Asia. When this great volume of ice melted, enough water was added to the ocean to raise global sea level by approximately 375 feet. The shoreline of 20,000 years ago was at the outer edge of the U.S. East Coast's continental shelf and moved westward as sea level rose. The initial rate of ice melting and sea level rise was relatively rapid but has slowed significantly over the last five thousand years. Now there are tide-gauge records and satellite measurements that indicate that the rate of sea level rise has begun to increase again.

South of the glaciated areas, sea level generally rose as the glaciers melted, but in the areas that had been covered by thick ice, like northern New England, the wasting of the glaciers affected local sea level in a different way. During the Ice Age, the great weight of glacial ice depressed the land surface and the crust of the earth beneath it. Thus, as the melting ice front retreated as far north as the Maine coast, the land was still depressed and the ocean flooded inland areas for more than 50 miles. Evidence for this drowning remains in the form of shoreline features now raised high above sea level at far inland locations. Ocean mud with 12,000- to 14,000-year-old fossils abounds in coastal lowlands north of Boston. Deltas and coastal landforms, above present sea level, have been observed in many places. Along the cliffed coasts of Acadia National Park in Maine, modern and ancient sea stacks, sea cliffs, and boulder beaches are within easy walking distance of one another. Sea cliffs and columnar sea stacks are common erosional features, cut by waves in solid rock, that are indicative of sea level position at the time they formed.

Following the removal of ice from these areas, the land rebounded, or rose, to its "normal" elevation and, as a result, local sea level fell. For example, as the glaciers melted off Maine, the land rose (sea level fell) 200 feet, resulting

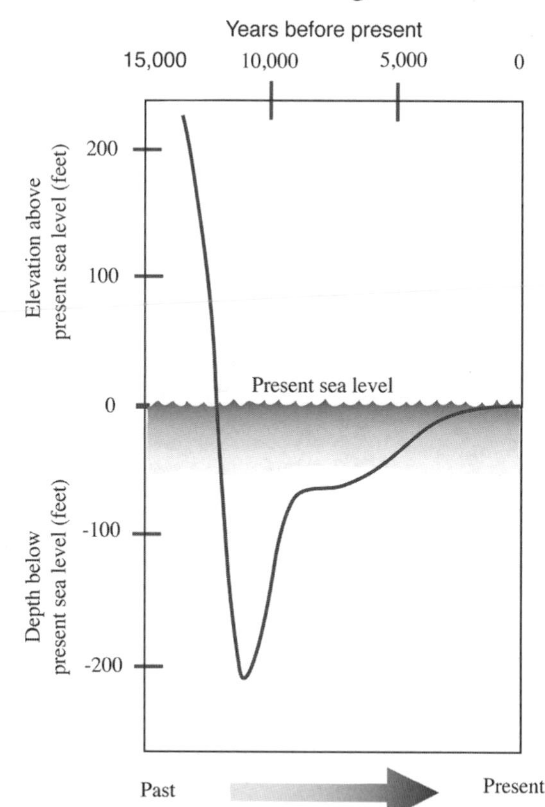


The sea level curve for Maine shows that sea level was relatively higher 15,000 years ago because the land was depressed by the weight of ice sheets. After the removal of the glacial ice mass, the land rose and sea level fell 200 feet below present level. Sea level has been rising for the last 10,000 years. —Drawing by Charles Pilkey

in a sea level fall and oceanward migration of the shoreline. Scientists from the University of Maine have mapped the old shorelines using sonar, submarines, and other techniques. This adjustment occurred about 12,000 years ago. Since then, land rebound has decreased and sea level has slowly risen in this area, although at varying rates. With rising sea level, though, the shoreline has migrated landward to its present position and continues to move inland.

How do we identify old shorelines and measure sea level changes? Generally, a geologist looks for something with a known relationship to the shoreline that is above or below modern sea level. The close proximity of modern and ancient features such as sea stacks in Maine, or coastal features such as dune ridges now found inland from Virginia to Florida, indicate how high sea level was in the past. Establishing the timing of sea level changes is much more difficult.

Geologists rely on radiocarbon dating of plants and animals that lived right at sea level, or close to the shoreline, to assign a time to past sea level locations. All living things acquire a radioactive form of carbon while they are alive. Once they die, the radioactive carbon decays at a known and regular rate. As time passes after the death of any organism, the proportion of radioactive ("unstable") to nonradioactive carbon ("stable") decreases. By measuring the amounts of stable and unstable carbon, a geologist can calculate how long a plant or animal fossil has been dead.

Seashells like *Mya arenaria* (the soft-shelled clam), which today live mostly between high and low tide, can be found in gravel pits above present sea level. In more southerly climes fossil oysters lived near sea level at or just below low tide, and their shells are commonly preserved where they grew in mud deposits of ancient lagoons. Cores collected from the seafloor sometimes contain unbroken *Mya* shells or oyster shells in the position in which they lived. In both cases, the shells of either species represent a former shoreline

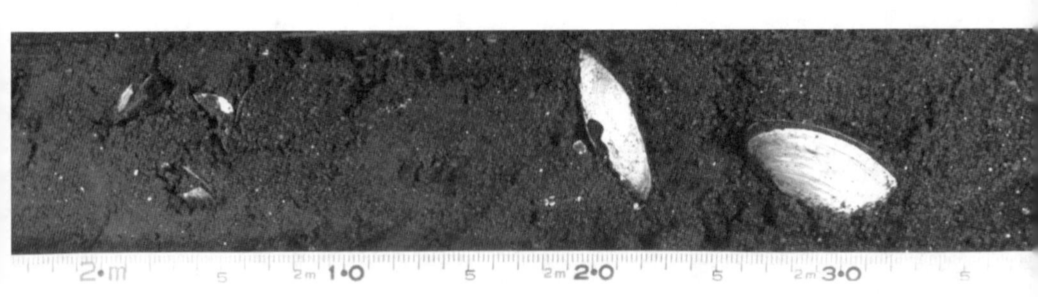


A core of seafloor sediments from 60 feet below present sea level. A clam, *Mya arenaria*, appears in the position it held when it was alive. This species of clam lives in the zone between high and low tide, so its occurrence in this core indicates that sea level was once lower where the core was taken.

A raised sea stack in Acadia National Park (Maine). An ancient sea cliff is left of the stack, and a fossil boulder beach lies at base of the stack under leaves. Many sea caves were cut into the sea cliff when sea level was higher.

A modern sea stack in Acadia National Park. Note the boulder beach at base of the stack and the sea cliff behind it. The boulders on the beach form as waves erode the bedrock sea cliff and sea stack.

As sea level rises and saltwater floods coastal forests, extensive areas of trees are killed. Salt marsh now grows around this former forest in the Wells National Estuarine Research Reserve (Maine).

position, either above or below present sea level. By radiocarbon dating the shells, one can assign a time before present when sea level was at that position. By plotting the ages and elevations of many of these samples, geologists can get an idea of how sea level has risen and fallen through time.

For the past few thousand years salt marshes have existed in relative abundance behind barrier beaches. These plant communities live near the high-water mark and grow upward and inland to keep pace with rising sea level. As the rising sea kills upland forests, salt marshes are established on top of the land that harbors the tree stumps. By radiocarbon dating the age of the first salt marsh plants that grew over the old forest floor, geologists can track the rise in sea level over the past 5,000 years. The slowing in the rate of sea level rise during the past 5,000 years all along the U.S. East Coast is interesting because it is believed to be the reason why barrier beaches and salt marshes have grown to the large proportions that we see today. The slow rate of sea level rise allowed abundant time for sand to accumulate in dunes and beaches, and for mud and peat to expand salt marshes across tidal flats. But, as noted above, the rate of sea level rise has accelerated in recent time.

Marsh peat overlying tree stumps provides a measure of how much sea level has risen at this location in the last 2,000 years.

The government established tide gauges along the East Coast to aid in the prediction of tidal heights. By the 1940s it became apparent that the ocean did not simply go up and down with the tides. Instead, when tracked over decades, sea level generally went up a bit more than it went down! The rate of sea level rise measured from tide gauges varies from place to place owing to continuing, small land-level adjustments. All along the East Coast, however, tide gauges have recorded a relatively rapid rise in sea level for the past century, and today satellite measurements also indicate that the sea is rising. At the research pier in Duck, North Carolina, scientists have measured sea level rise on the order of 1½ feet per century. In many places sea level has risen a foot or more since 1900, whereas 1,000 years ago it might have risen only an inch or so over the same time span. Although a foot or so of sea level rise in one hundred years might not seem like much, the response of beaches to this rise can be pretty extreme!

## Barrier Islands: Hot Dogs, Drumsticks, and Broken Spits

A barrier island is a special sort of island. Unlike the rocky islands of New England, a barrier island is mostly a pile of sand, but it behaves like a living

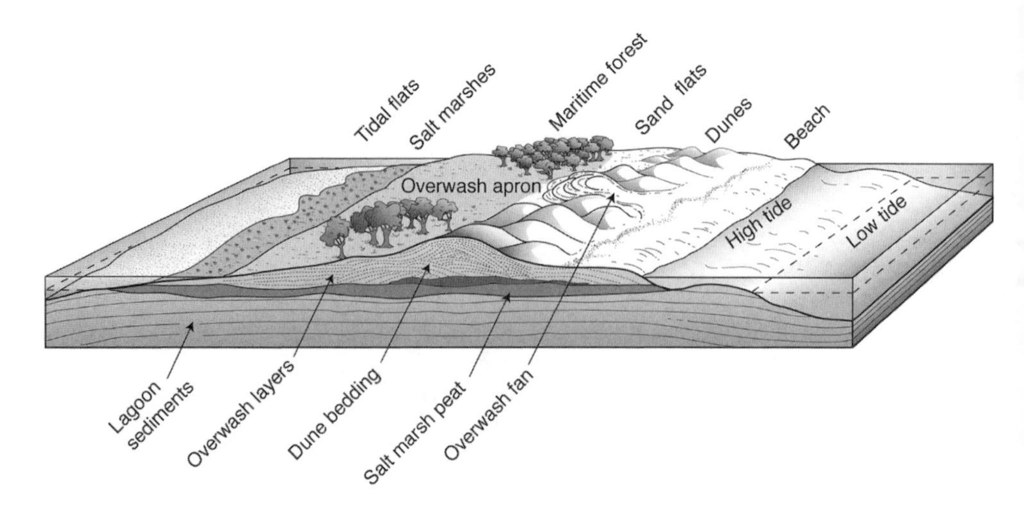


A typical barrier island. The beach represents the high-energy ocean side of the island and is the sand source for the dunes. Sand dunes form at the back of the beach atop earlier overwash deposits. Continued overwash may carry beach sand into the dunes or across the island into the lagoon on the back of the island. As adjacent overwash fans coalesce, they form an overwash apron, also called an *overwash terrace*, atop lagoonal, tidal flat, and salt marsh sediments (mud and peat). The transport of sand from the beach to the lagoon by overwash and dune migration is the mechanism by which a barrier island migrates landward. —Drawing by Charles Pilkey

pile of sand. Barrier islands are longer than they are wide, typically miles in length and a fraction of a mile wide, and they have an ocean on one side and a lagoon on the other. Separating islands are inlets, through which the tides exchange estuarine water (including freshwater runoff from the land) with the ocean. On the seaward side of barrier islands, the shoreface dips relatively steeply from the shoreline to a depth of 30 to 60 feet, usually less than a mile from the beach. Beyond the shoreface, the much more gently sloping continental shelf begins. It is easy to distinguish between the flat continental shelf and the relatively steep shoreface on navigation charts since depths change rapidly across steep slopes.

More than 2,500 barrier islands around the world make up about 10 percent of the global open-ocean shoreline. They range in length from a few hundred yards to the 135-mile-long Padre Island (Texas); however, the average global length of barrier islands is on the order of 10 miles. If one includes the Alaskan barriers above the Arctic Circle, a full 25 percent (by length) of the world's barrier islands, 405 in number, are owned by the United States. With a few breaks, the barrier island chain in the lower forty-eight states extends all the way from just south of Montauk Point, at the tip of Long Island, to the Mexican border. More than 75 percent of the East Coast shoreline is lined with barrier islands and their beaches.

No two barrier islands are alike for the same reasons that no two beaches are exactly alike. Each beach and each island has a unique combination of orientation, sand supply, sand type, vegetation, winds, waves, tidal range, storms, and offshore topography. The underlying geology of barrier islands differs as well and often contributes to island evolution. For example, Miami Beach, Florida, sits on an ancient coral ridge. Those corals formed perhaps 120,000 years ago and are now covered with a layer of sand that is often no more than 3 or 4 feet thick. The same is true for Palm Beach and for parts of Fort Lauderdale, Florida. At Palm Beach, the old coral ridge is visible in some road cuts.

A few stretches of mainland disrupt the continuity of the barrier island chain along the East Coast. For example, from Bay Head north to Monmouth Beach, New Jersey, there is a 20-mile long stretch of mainland headland. Before this entire shoreline stretch was locked up by seawalls, the erosion of the headland provided sand to the Sandy Hook spit to the north as well as the barrier islands to the south. Other mainland shoreline reaches include a portion of the Delaware coast; Virginia Beach, Virginia; Kure Beach, North Carolina; Myrtle Beach, South Carolina; and parts of the Florida coast. These are coastal stretches where the migrating barrier islands have virtually caught up with the mainland.

Rising sea level and an irregular coastline are two conditions that contributed to the origin of barrier islands. The shoreline of the Coastal Plain was mostly straight and regular when sea level was lower and streams dumped their sand load directly into the ocean. As sea level rose, river valleys were flooded, forming estuaries and embayments that were separated by headlands. These headlands that extended seaward were the ridges or divides that had separated the river valleys. As sea level continued to rise, waves eroded the protruding headlands, providing sand to spits that were developing off the headlands. The flooded valleys behind the spits became sounds and lagoons. Eventually, storms broke through the spits, forming inlets, and the barrier islands were born.

Barrier islands are still forming today. At Chatham, Massachusetts, on Cape Cod, a new island formed as an inlet cut across Nauset Spit during a storm in 1987. Historic navigation charts and maps show that this inlet-cutting process has occurred several times since European colonists arrived. This system opens and closes with a periodicity of about 150 years, illustrating the ever-changing nature of barrier islands. A new inlet formed across the Outer Banks of North Carolina during Hurricane Isabel, in 2003, but was quickly filled in by dredges and bulldozers. The 1938 hurricane created a barrier island at Napatree Spit in Rhode Island. The new barrier island has now migrated into Connecticut waters.

Besides rising sea level, which is needed to form an irregular coastline, barrier islands also require a gently sloping upland of a regional scale immediately adjacent to the shoreline, a large supply of sand, and sufficiently energetic waves to move the sand about. These conditions are required in order for barrier islands to form as well as continue their evolution. A steep upland, like that which occurs in much of New England, prevents back-barrier bays from forming, so a migrating barrier island becomes a mainland beach when it reaches this upland. Without sand and waves, beaches cannot exist at all.

Barrier islands tend to evolve toward a state of balance, or equilibrium, adjusting to changes in wave-height conditions, sea level, and the supply of sand. Because one or more of these controlling factors is changing almost continuously, the perfect equilibrium is never fully reached. Once formed, the islands have the marvelous ability to migrate in a landward direction as sea level rises and waves and winds drive sand landward from the shoreface and continental shelf. But where the sand supply is too low or the rate of

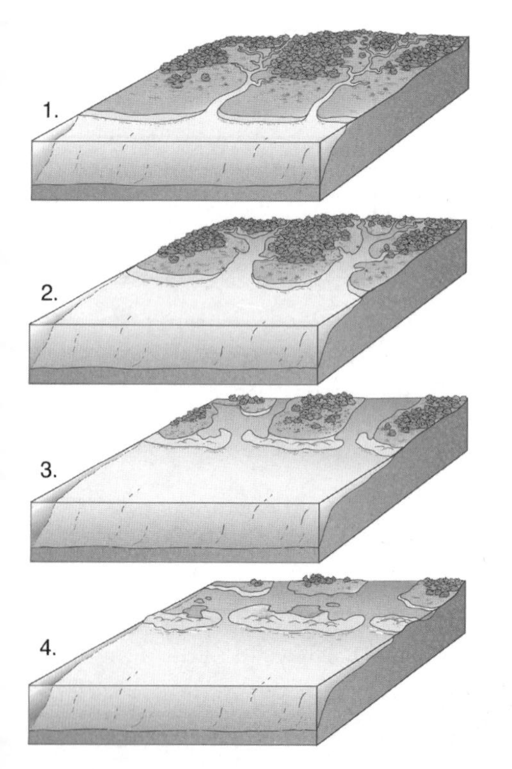


The formation of barrier islands as sea level rises. Stage 1: Straight coastline forms during lower sea level. Stage 2: Sea level rises and floods valleys, forming a sinuous coast. Stage 3: Sand eroded from resulting headlands forms spits. Stage 4: A combination of breaching by storms and flooding of low-lying areas behind spits forms the barrier islands. —Drawing by Charles Pilkey

sea level rise is rapid, as on the edge of the subsiding Mississippi Delta in Louisiana, barrier islands drown. When barrier islands do move landward due to a sea level rise, the mainland shorelines also retreat as they are flooded by the same sea level rise. And simultaneously, an island's lagoonal or sound shoreline builds out into the backwater.

Storms are the events that drive island migration. Storms cause the beach to retreat on the ocean side and also widen the island. Widening results when sand is taken from the beach and washed across the island to be deposited in the lagoon. The bodies of storm sand deposited across an island are called *overwash fans*. Such fans can be most clearly seen on undeveloped islands, such as those along Virginia's portion of the Delmarva Peninsula, and on the Cape Lookout National Seashore in North Carolina. To identify overwash deposits on barrier islands, look for large seashells and coarse shell debris well inland from the beach. Overwash that has reached the back side of an island often appears as fan-shaped lobes that bury the salt marsh. These overwash fans often occur landward of gaps in frontal dunes, where elevated storm surges can most easily overtop the dunes and gain access to the back side of an island. Such breaches in the dunes are often the result of

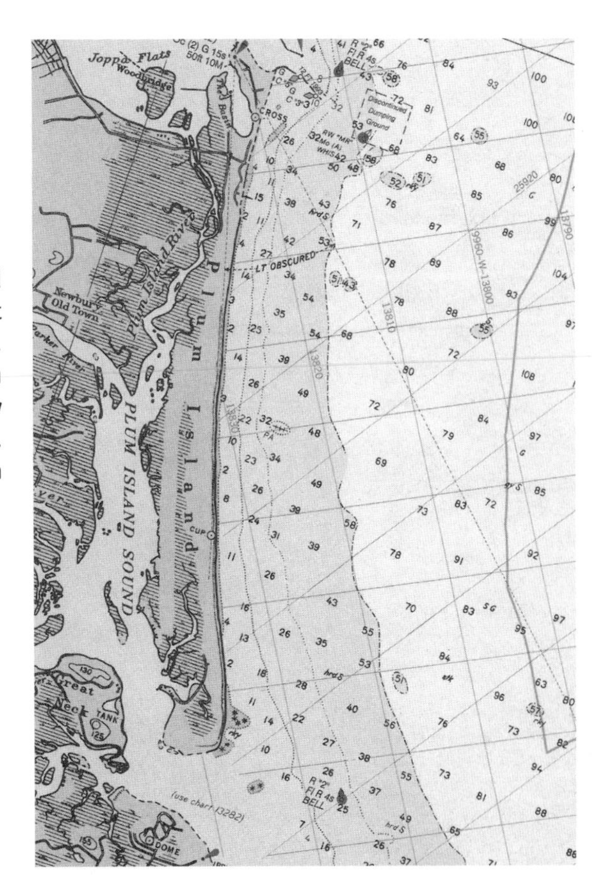


Nautical chart of Plum Island (Massachusetts), the northernmost barrier island in the United States. The island's beach continues beneath the ocean surface as a sedimentary slope to as far as 3 miles offshore. —Courtesy of the NOAA, National Ocean Service, Coast and Geodetic Survey

The back side of Plum Island (Massachusetts) is a series of lobes, which were produced by storm overwash as well as the incorporation of flood-tidal deltas when old inlets closed. Sand moving through inlets or across the island, for example, as overwash and sand dunes, allows the island to migrate landward as sea level rises.

disturbances to natural vegetation from vehicles, footpaths, and fire that destabilize the sand.

On developed islands, the classic fan-shaped overwash deposit may not be as apparent after storms, but overwash sediment is the material that blocks streets and driveways and fills in swimming pools and the ground-level floors of buildings. Such sediment is often hauled away during poststorm cleanup, but it should be returned to the beach and sand dunes as it is critical to beach-dune equilibrium.

Many of the East Coast barrier islands probably formed well seaward of where they are today and migrated to their present locations over the last several thousand years. Most barrier islands are not presently migrating because they are covered with buildings, their sediment sources and processes are blocked by engineered structures such as seawalls and groins, and their beaches are being held in place through artificial beach nourishment. The overall slowing of sea level rise in the past few thousand years also allowed sand dunes to build and inhibit overwash that prevailed during earlier times of more rapid sea level rise. But sea level has begun rising again, starting about 150 years ago.

Migrating islands include Assateague Island (Maryland and Virginia); Masonboro Island (North Carolina); several islands at Cape Romain (South Carolina); and Cabretta Island, part of Sapelo Island (Georgia). In New England,

Overwash fans on Assateague Island. During storms, waves carry beach sand across the island and deposit the sediment as lobes, which extend onto the salt marsh. The net effect is that the island is rolling over itself, migrating landward (to the left). Ocean City, Maryland, is at top of photo.

Nauset Spit on Cape Cod and Monomoy Island have also migrated landward. The migrating island with the easiest public access is the Assateague Island National Seashore, south of Ocean City, Maryland. The migration in this case can be directly attributed to the jetties at the north end of the island (the south end of Ocean City, Maryland) that have trapped sand from longshore currents—the sustenance of all barrier islands—and starved Assateague Island. Deprived of sand, the ocean shoreline of northern Assateague Island has migrated behind what was its lagoon shoreline in the 1930s. Masonboro Island (North Carolina) is also starved of sand by jetties and is moving landward at the rapid rate of 16 feet per year.

The presence of salt marsh mud and peat on an open-ocean barrier island beach is a sure sign that an island has migrated. Islands in Virginia, the Carolinas, and Georgia have mud and peat layers on their beaches. These muds formed in salt marshes that once existed on the back sides of the islands, before the islands migrated landward, over the marshes.

Mud is almost always visible on Cedar Island (Virginia); Edisto Island (South Carolina); and Wassaw, Ossabaw, and Sapelo Islands (Georgia). Dozens of other shorelines, including some in New Jersey, Maine, and Massachusetts, may briefly exhibit mud and peat layers on beaches after storms. Generally, sand covers the mud patches as the beach recovers from the storm.

Mud exposed on the beach on Sea Island (Georgia) in the 1980s. The beach had steepened because waves were reflected off the seawall and the seawall cut the beach off from its sand supply. As a result, the underlying mud, which contained roots of salt marsh grass, was exposed.

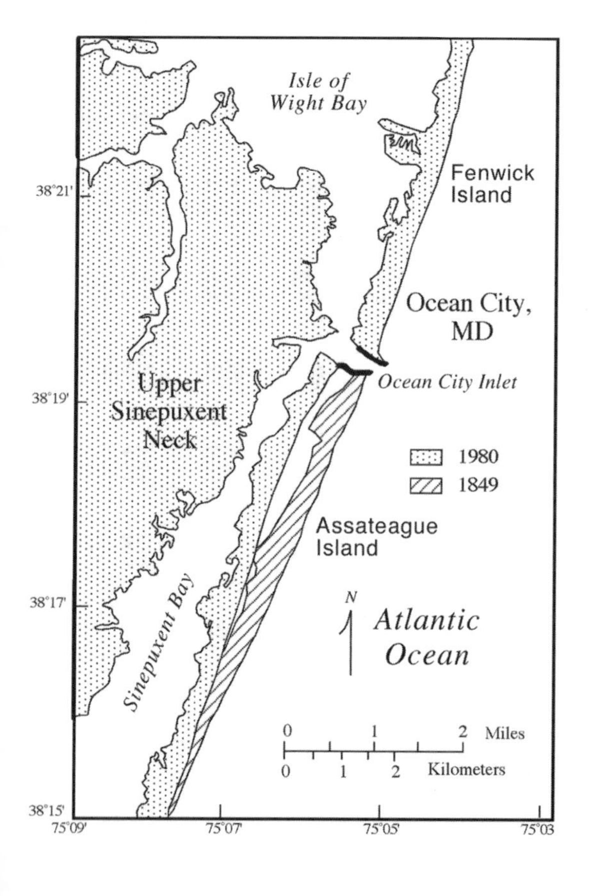


Map of Assateague Island showing the 1849 position of the island and the post-jetty location as of 1980. Ocean City Inlet opened during a storm in the 1930s, and the jetty was constructed to keep the inlet open for navigation. The jetty blocked the north-to-south flow of sand, cutting off the sediment supply to Assateague Island. The island migrated a distance greater than its width in the time interval from the late 1930s to late 1970s. —Modified after Williams, S. J., et al (1991)

Tree stumps are another common feature on beaches, especially after storms. These stumps indicate beach erosion or shoreline retreat, but they do not indicate that island migration has occurred. That is, the trees come from the middle of the island and not the lagoon behind the island where the marsh mud and peat come from. Beach retreat into the maritime forest results in a maze of fallen trees on the beach. In this case, the distinction between island migration and shoreline erosion becomes clear. The island does not have to migrate for erosion to occur, and fallen trees on a beach are clear evidence of beach erosion. Good examples of trees on barrier beaches include Wassaw and Jekyll Islands (Georgia) and Hunting Island (South Carolina). It is important to emphasize the difference between island migration and shoreline erosion, or retreat. Migration refers to the movement of the entire island in a landward direction. Shoreline erosion refers only to the landward movement of the ocean-facing beach. Shoreline erosion is one of the processes that is involved in barrier island migration, but the two natural events are separate processes.

Marsh mud and peat are not the only evidence that islands have migrated inland in the past. Many of the seashells on a beach were derived from the lagoon environment on the back side of the island rather than the beach and the continental shelf on the ocean side. Oysters, which live in sheltered estuarine waters, are by far the most common lagoonal shells that can now be found on beaches.

Tidal deltas that form at inlets are also integral parts of barrier islands. Tidal deltas are the bodies of sand that tidal currents have built out into the open ocean or into a lagoon. They have caused—and will continue to cause—many a vessel, large and small, to come to grief as it tries to enter the harbor behind a barrier island. Deltas built into the open ocean are called *ebb tidal deltas*, and those built on the back side of a barrier island are called *flood tidal deltas*. The size of these sand bodies varies greatly, depending on the magnitude of local waves and tidal currents.

For example, off St. Simons Island (Georgia) the ebb tidal deltas are huge, extending as much as 10 miles offshore. Such large ebb tidal deltas occur where ocean waves are relatively small but the tidal range, also known as tidal amplitude, is large (7 to 11 feet between high and low tide). The large tidal range, the vertical difference between normal high and low tides, causes strong ebb currents to move sand seaward on the falling tides. The small waves are not able to disperse this sand, so it builds into a delta. Low waves and high tides produce large *ebb* tidal deltas but small *flood* tidal deltas. Higher tidal ranges force larger volumes of water into and out of the

These stumps were exposed at milepost 13 in Nags Head, North Carolina, by a mild September storm in 2005 that arrived from the southeast. —Photo by Ray Midgett

The same beach had recovered three days later after the storm winds had stopped and fair-weather waves had accreted sand onto the beach, reburying the stumps. Local beach users have observed that such stump exposure has occurred repeatedly, and almost always during storms with winds blowing from the southeast. —Photo by Ray Midgett

Tidal deltas are apparent in this aerial photo of Drum Inlet on the Outer Banks of North Carolina. The ebb tidal delta on the seaward side of the inlet is relatively small in comparison to the large flood tidal delta in the lagoon. When an inlet closes, the flood tidal delta becomes a platform on which a new salt marsh develops. This is how the lobate marsh *behind* the island and above the flood tidal delta formed.

lagoons and estuaries, resulting in larger, more closely spaced inlets and shorter barrier islands.

Off Core Banks (North Carolina), a string of barrier islands that stretches from Ocracoke Inlet to Cape Lookout, the ebb tidal delta at Drum Inlet in Cape Lookout National Seashore extends only a mile or so offshore, but the flood tidal delta is much larger than the flood tidal deltas in South Carolina and Georgia. Here, much larger ocean waves break on the outer rim of the ebb tidal delta and drive sand close to shore. At the same time, the lower tidal amplitude produces smaller tidal currents, but wave and tidal currents flush sand through the inlet during storms and at high tide to form a delta on the lagoon side of the barrier island. High waves and low tidal amplitudes produce relatively small *ebb* tidal deltas and large *flood* tidal deltas. Lower tidal amplitudes also result in fewer, more widely spaced inlets and longer, skinnier barrier islands.

Perhaps the most important aspect of ebb tidal deltas as far as barrier island beaches are concerned is their impact on waves that strike the beach. The large body of sand that extends seaward of the inlet causes waves approaching the shore to refract, or bend, causing an uneven distribution of wave energy along the beach. As a result, on some islands more sand is deposited along the island front near its inlets, widening the ends of the island. On the East Coast, powerful storm waves generally come out of the northeast. Thus the highest wave energy strikes the northeast-facing margin of the ebb tidal delta, sheltering the *north* end of the next down-drift island while the rest of the island's shore is eroding, producing a narrower or skinny island. In map view, these skinny islands with bulbous ends look like chicken drumsticks. Drumstick-shaped islands are particularly well developed in South Carolina because the ebb tidal deltas are large, for example, at Folly Island, Isle of Palms, and Fripp Island. On the East Coast the wide part of the barrier island drumsticks are usually at the north end of the islands.

The opposite of a knobby drumstick barrier island is one that is long, straight, and narrow. As noted, where the tidal range is low (less than 6 feet), inlets are widely spaced with fewer and smaller ebb deltas. With less wave refraction, the longshore transport of sediment is the dominant process, so these islands have a more uniform width along their entire length, giving

Drumstick-shaped barrier islands form where wave energy is lower and tides occur in the mid-tidal range (6 to 13 feet), resulting in more closely spaced inlets. Because of the relatively large ebb tidal deltas, more sand is deposited near the inlets, fattening the end of the drumstick. —Photo by Andy Coburn

rise to their designation as *hot dog islands*. The barrier islands of the Outer Banks (North Carolina) are examples of well-developed hot dog islands. (The hot dog–drumstick island classification is the brainchild of Miles Hayes, a pioneering barrier-island geologist, formerly at the University of South Carolina.)

Georgia's barrier islands are a special type; they are called the *Sea Islands*, which is both a geologic and geographic term for these islands. This string of islands spans the coast from Hilton Head Island in South Carolina to Amelia Island in northern Florida. The older islands are a combination of modern islands, a few hundred years old, that welded, or are in the process of welding, onto much older islands that formed during periods of higher sea level during the Ice Age. The Sea Islands are much larger than the younger, modern barrier islands seaward of them. They are probably 12,000 years old. The younger islands are migrating toward the larger islands and are, or will soon be, incorporated into them. St. Simons Island (Georgia) is an ancient island, and the Sea Island resort is on the modern island in front of it. Sapelo Island, also in Georgia, is the ancient island, while Cabretta and Blackbeard Islands are the associated modern islands.

Florida's Atlantic Coast islands are yet another island "breed." Only the northernmost two Florida islands are true natural barrier islands. Before the twentieth century, the rest of the Florida shore was a long sand ridge that didn't have islands and was only broken by a few river mouths. During the early part of the twentieth century, a number of channels were dug or blasted across the sand ridge, admitting saltwater into what was once mainly freshwater marsh or river channels behind the ridge, creating islands. Digging the intracoastal waterway, the Atlantic coastal navigational system, behind the old sand ridge completed the job of lining the Florida Coast with islands. Salt marshes and mangrove forests arrived on the east coast of Florida only after the artificial inlets were constructed.

In New England there are few barrier islands because the mainland is high and rocky rather than being a flat, sandy Coastal Plain. Nevertheless, there are barrier islands, but people have connected most of them to the mainland and other beaches by filling inlets and wetlands. Drakes Island, in Wells, Maine, is perched atop a ridge called a *moraine*, which was deposited at the front margin of a glacier as it melted. Erosion of the glacial material provided sand and gravel for this island's beach until most of the eroding sand supply was artificially walled up to protect property. Plum Island (Massachusetts) is another good example of a New England barrier island that gets part of its sand supply from an eroding glacial deposit; the rest comes from a nearby river and the continental shelf.

Erosion of a glacial deposit (drumlin) on Plum Island National Seashore (Massachusetts) provides the sediment supply to the beach. This mixture consists of sand and gravel with particles as big as boulders.

Eroding glacial till (moraine) at Drakes Island near Wells, Maine, results in cobble beaches as smaller sediments are carried away and the cobbles are left behind.

If there were no building and development on the East Coast barrier islands their future would be an exciting one. The islands would gradually narrow in response to the present sea level rise. This narrowing is happening to the islands of the Cape Lookout National Seashore (North Carolina) and some off the Delmarva Peninsula. Once the islands become narrow, storms will routinely transport sand across the islands or new inlets will cut across them and sand will be carried to the lagoon side of the island, allowing the islands to migrate landward.

True island migration, such as what is happening with Assateague Island or Masonboro Island (North Carolina), will probably begin for most East Coast barrier islands in the next one or two centuries. Of course, this natural scenario is not going to happen if people can stop it. Instead, for future generations the barrier islands will become a battleground. People will continue to try to out-engineer nature, even as sea level rises and the islands move. Humans want the islands to hold still!

The narrowing and lowering of barrier islands allows new inlets to form during storms, as seen here on Hatteras Island (North Carolina) during Hurricane Isabel in 2003. Sand is carried through the inlet to the back side of the island to form a new flood tidal delta. Overwash fans are also apparent on the back side of the island (left). This inlet was closed by the Army Corps of Engineers shortly after it opened. This inlet also opened at the same location during a previous storm but soon closed due to natural processes.
—Photo by © Sidney Maddock

## Beach Erosion and Accretion: Moving Beaches

If sea level is rising and the shoreline has moved landward from roughly the 375-foot water depth along the Atlantic Coast, then where on the beach is there evidence of this movement? Although long-term visitors to a given beach have probably noticed changes over the years, especially after storms, undeveloped, natural beaches look much the same from year to year. Beaches do not generally move landward during sunny summer days when most beach visitors are frolicking in the sand. Beaches move during storms, especially big storms.

Nauset Light Beach on Cape Cod National Seashore (Massachusetts) illustrates what happens to a beach during a really large storm. Before the winter of 1978, a parking lot and bathhouses covered much of this barrier spit behind its frontal sand dune. A barrier spit is the precursor of a barrier island; one end is still connected to the mainland. Northeast winter storms in January and February of 1978 assaulted the beach with fury. The bathhouses were destroyed, waves swept across the dunes and parking lot, and a large volume of sand was carried into the lagoon behind the beach. The National Park Service wisely decided not to rebuild either the bathhouses or parking lot. When we visited the beach in 1985 the former parking lot was a sand flat with chunks of pavement scattered about.

Geologists call such a landform an *overwash flat*, or *overwash apron*, because ocean water and sand wash over it during large storms. During a storm, sand is also moved from the beach out into the bay, and the loss of sand usually causes the shoreline to move landward. The overwashing of barrier spits and islands is a major mechanism that facilitates the landward migration of beaches and barrier islands. By 2002 at Nauset Light Beach, dunes had already begun to cover the overwash flat. Another century may pass at Nauset Light Beach before a big storm again washes over it, or it could happen tomorrow, but just as surely as the ocean is rising, the storm will come. In contrast, low-lying Masonboro Island (North Carolina) overwashes several times each year.

The best evidence for the landward movement of beaches is found on the beach itself. Often following a storm, salt marsh peat deposits and tree stumps appear in the intertidal zone. The peat and stumps represent environments that existed behind the beach, so their appearance indicates landward beach movement over the forest and marsh. Examples of such landward-moving islands include Little Talbot Island (Florida); Blackbeard and Jekyll Islands (Georgia); Hunting Island (South Carolina); Cedar Island (Virginia) and Assateague Island (Maryland and Virginia); Cape Henlopen (Delaware); Brigantine and

The parking lot and bathhouses at Nauset Light Beach at Cape Cod National Seashore (Massachusetts) were separated from the beach by a frontal dune in 1978. —Photo courtesy of the National Park Service

By the 1980s the same area shown in the photo above was a overwash flat. Chunks of broken pavement, visible in the photo, were washed onto the sand flat.

By 2002 dunes were forming on and covering the overwash flat.

Peat was exposed on Folly Island (South Carolina) in the 1980s, providing evidence that the beach had moved landward.

Ludlam Beaches (New Jersey); Napatree Spit (Rhode Island); First Encounter Beach in Cape Cod Bay, Slocum River embayment in Buzzards Bay, and Crane Beach at Castle Neck (Massachusetts); and Drakes Island in Wells, and Jasper Beach at Machiasport, Maine, to name just a few. Studies of core samples collected from beaches and barrier islands across the East Coast also show the connection between living marshes, "dead" peat deposits outcropping on the beach, and beach migration. These cross sections indicate that barrier islands migrate landward across back-barrier deposits like a tank tread. In this way barrier islands can survive drowning as sea level rises.

The landward movement of coastal environments in response to sea level rise is termed *transgression*. Examples of transgression include a dune field migrating landward and burying a salt marsh, or the salt marsh encroaching on the adjacent mainland forest. In each case a "more landward" environment is replaced by one that existed in a "more seaward" location.

Historic maps also show landward shifts in the position of barrier islands. Napatree Spit, originally in Rhode Island and now partly in Connecticut, was one of the first beaches where the effects of a large storm were documented. Prior to 1938, Napatree Spit was attached to the mainland near Watch Hill, Rhode Island. The Great New England Hurricane of 1938 separated the spit from the mainland (with considerable property damage), forming a barrier island, and overwash has led the barrier to migrate landward. At some point the island crossed into Connecticut, where it continues its journey!

A map depicting shoreline change at Napatree Spit (formerly in Rhode Island, now in Connecticut). The great New England Hurricane of 1938 separated the spit from the mainland, and the new barrier island migrated landward into Connecticut waters. —Modified from a handout compiled by Dr. John Fisher

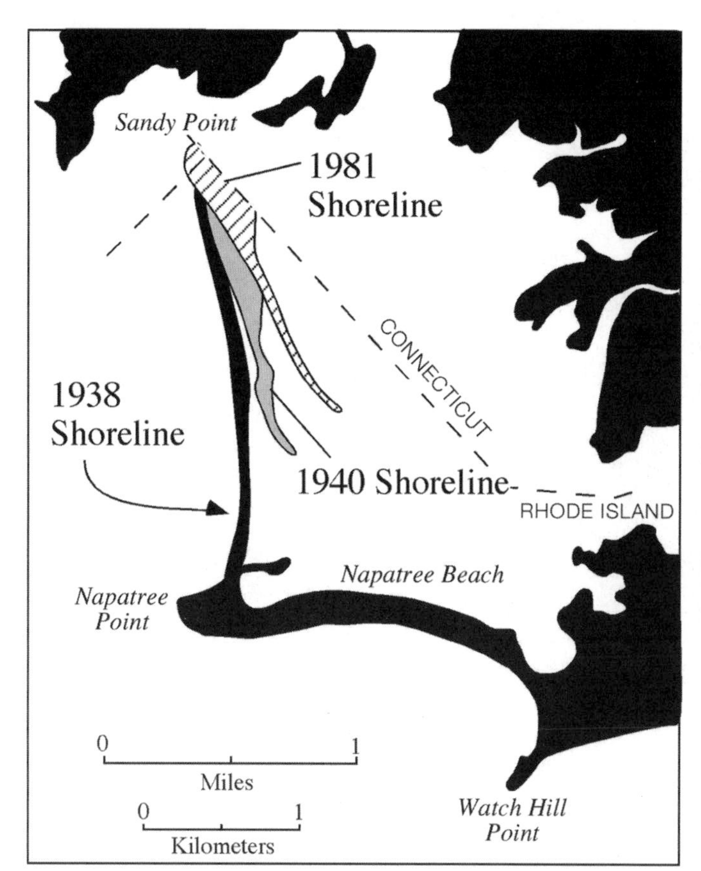

A 2005 aerial view of the barrier island that formed after Napatree Spit separated from the mainland in 1938.

The Morris Island Lighthouse (South Carolina) was left with its feet in the water after Morris Island migrated landward. The lighthouse was still on land in the 1940s and is now about 2,000 feet offshore, illustrating a rapid rate of shoreline retreat.

Not all barrier island movements are caused strictly by sea level rise. When the supply of sand to a beach is interrupted, the beach behaves in a manner similar to a sea-level-driven transgression. In the absence of sand, waves attack and breach dunes, causing overwash events. A classic example occurred at Assateague Island National Seashore. Construction of a jetty in 1933 at Ocean City, Maryland, cut off the island's longshore sand supply. Since then, Assateague Island has moved landward a distance greater than its original width.

Morris Island (South Carolina) provides another example of rapid island transgression due to human interference with the island's sand supply. Jetties were constructed a few miles north of Morris Island in 1896 to protect the entrance to Charleston Harbor. Unfortunately, the jetties also blocked the north-to-south sediment transport that supplied Morris Island with sand, so the island began its rapid migration. The dune ridges migrated landward, as did the beach, particularly at the southern end of the island. The Morris Island Lighthouse, constructed between 1874 and 1876 in the marsh about 600 feet behind the 1876 beach, now stands a good 2,000 feet offshore with its feet in the sea.

The landward movement of barrier islands and shorelines is not the only way a beach responds to rising sea level. In New England's glaciated terrain,

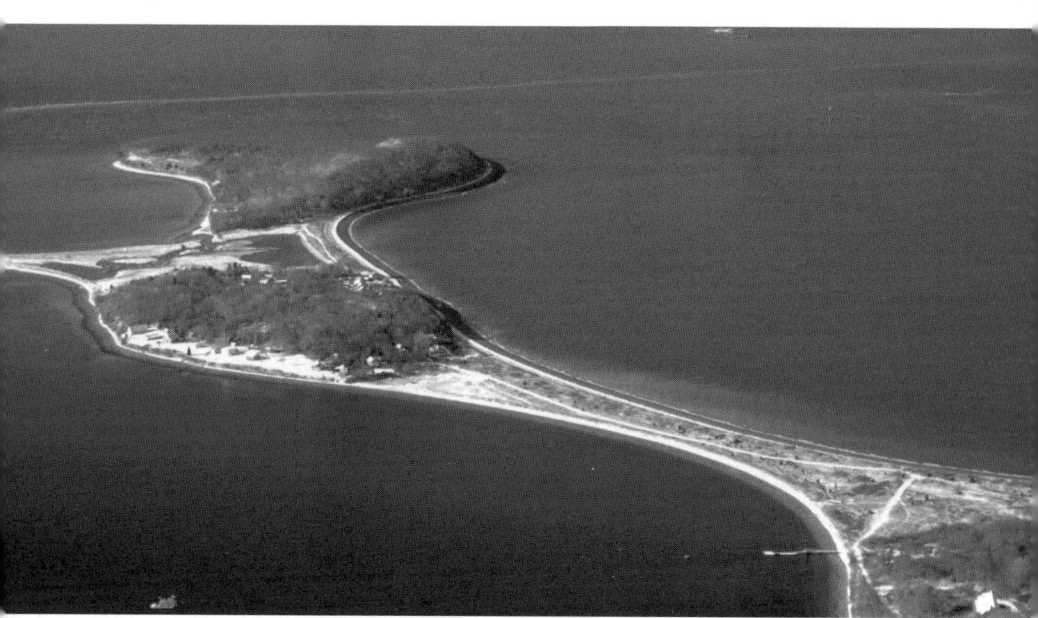

Boston Harbor's islands are partially submerged drumlins (small hills of glacial deposits). The erosion of these islands, not the continental shelf, supplies the sediment for the beaches.

when the supply of sand from an eroding bluff gives out, the shoreline moves back to a new bluff; sometimes, though, shorelines can actually drown in place. Boston Harbor's islands are all largely glacial deposits of sand and gravel that originally were drumlins, mounds or hills of sediment deposited and shaped by glaciers. Their beaches have moved inland for thousands of years from one bluff to the next. Where deep water separated one drumlin from another, beaches drowned when the glacial material was exhausted. Eventually, these islands would disappear if not for human intervention. Geologists have mapped drowned beaches and eroded bluffs in the offshore areas of New England by means of geophysical remote-sensing devices. Beach deposits below the sea and ramped up against bedrock cliffs are common offshore features in New England. These beaches migrated landward against a bedrock cliff, were partly dispersed by waves, and then drowned.

In a few locations, beaches are not migrating landward or eroding at all; instead, they are growing seaward. This seaward growth phenomenon is known as *accretion*, or *progradation*. Accretion occurs where more sand is added to a beach than is needed to counteract the effect of rising sea level and storms.

Accreting beaches are not too common. At Popham Beach State Park (Maine), sand supplied by the Kennebec River has fought off the ocean's erosional tendencies for a long time. Evidence for the expansion of the beach exists in the substantial back-barrier maritime forest. This forest has grown seaward, as shown by the ages of trees at the back dune, which are older, and frontal dune areas have expanded from the addition of sand.

A few beaches along the East Coast barrier islands are building seaward, but these are the exception rather than the rule. One of the best examples is Sunset Beach at the North Carolina–South Carolina border. The beach has been growing seaward at a rate of 2 or more feet per year for a couple of decades because of a large sand supply, but in this time of rising sea level, most geologists expect the beach to reverse direction before long.

Many East Coast shorelines on barrier islands were building seaward in the past. Barrier islands such as Bogue Banks (North Carolina), Kiawah Island (South Carolina), Jekyll Island (Georgia), and Amelia Island (Florida) have multiple rows of long dune ridges parallel to the shoreline called *beach ridges*. Each of these beach ridges represents a former frontal dune ridge that was once adjacent to the beach. This succession of ridges developed over the last few thousand years as the shoreline was accreting during a period of relatively slow sea level rise. Now the shorelines of these islands are eroding and sea level is rising more rapidly.

Most shoreline accretion is probably caused by people. Just as the jetties at Charleston Harbor were responsible for down-drift loss of sand and the rapid erosion of Morris Island, they also trapped sand on their updrift, or north, side at Sullivans Island, causing the south end of that island to prograde hundreds of yards. In general, the updrift side of jetties is an area of sand entrapment, where a beach builds out into the ocean. The artificial replenishment of beaches is also a form of short-term accretion, but it is short-lived because such projects are carried out on retreating beaches. The natural conditions of these beaches remain erosional, and the ocean removes the added sand and returns the shoreline to conditions of retreat.

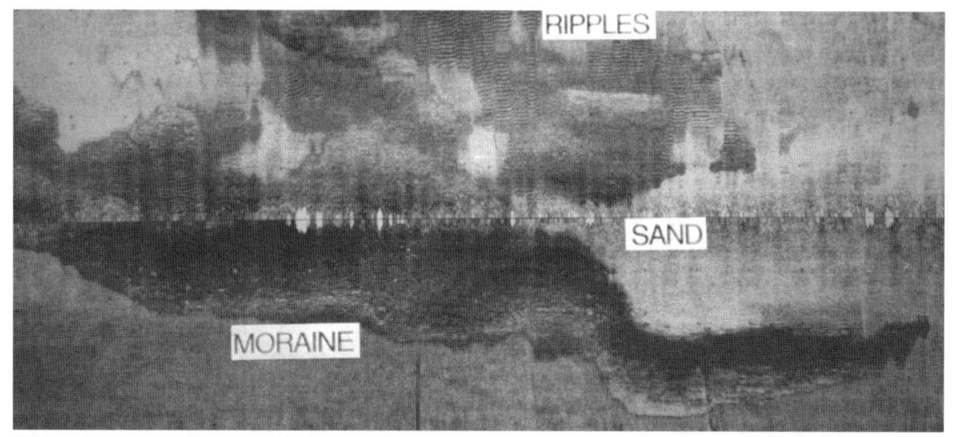


Side-scan sonar images of the seafloor are analogous to air photos of land. The sonar images provide a method for mapping old drowned coastal features such as beaches and moraines. In this image of the seafloor off Drakes Island, near Wells, Maine, the dark area is a moraine, while the lighter areas are sand bottoms, partly old beaches. Note that extensive areas of the sand field have large, wavy bed forms known as *megaripples*, which represent movement of sand by storm waves.

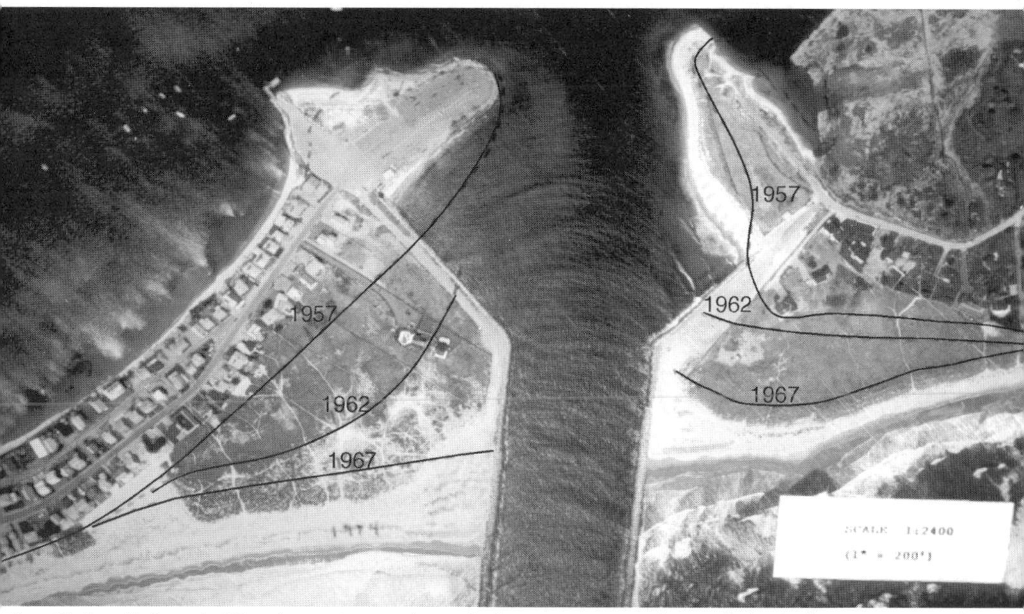


The Wells, Maine, jetties have trapped sand that once moved parallel to the coast, causing shoreline accretion. The lines represent past shoreline positions, which indicate that most of the land area shown has formed since the 1950s. The open-ocean beach faces the bottom of this photo. —Photo courtesy of the Maine Geological Survey

# Oceans In Motion

Having considered the framework in which beaches form, the variable timescale of different beach features, and the response of beaches to sea level change—the process most fundamental in shaping them over time—let us now turn to the forces that shape the beach on a daily to annual basis. Tides, waves, and nearshore currents constantly move sand and alter the appearance of beaches, sometimes supplying sand for beach growth or removing sand, causing retreat. The greatest of these forces are storm waves that occur during hurricanes and nor'easters, but day-to-day tides, ordinary waves, and longshore currents work constantly to produce the overall beach form and features we discuss in later sections.

## Perfect Storms: Hurricanes and Nor'easters

The Halloween Storm of 1991, later called the Perfect Storm, started out as a small, eastward-moving, low-pressure system over the Great Lakes. This small disturbance collided with an icy cold, low-pressure zone that was drifting down from Nova Scotia. This duo created a big storm that, in turn, collided with Hurricane Grace, a late-season storm that had plowed much farther north than most hurricanes do. The combined storms lasted from October 26 to 31, eventually producing 100-mile-per-hour winds out at sea, 100-foot waves in the open ocean, and 30-foot waves on some New England beaches. From New Jersey to Maine, approximately two hundred houses were smashed up, many properties flooded, and large changes occurred in some beaches.

Farther south on the southern Virginia beaches and on the North Carolina Outer Banks, huge, evenly spaced swells rolled ashore, flooding the seawardmost portions of the town of Nags Head but causing relatively little shoreline erosion. In fact, as often happens, swells from distant storms bring sand to the beach from offshore rather than eroding sand away. As the big waves rolled ashore from the Perfect Storm, wind velocities of less than 20 miles per hour were recorded. The winds responsible for forming these waves had been hundreds of miles to the north. Likewise, the swells from Hurricane Isabel in 2003, which passed well offshore of South Carolina, brought sand to Myrtle Beach, South Carolina, and did not cause erosion there.

The Perfect Storm was the largest nor'easter on record, exceeding even the famous Ash Wednesday Storm of 1962. Nor'easters derive their name from the counterclockwise flow of air around the storm's low-pressure center, which brings winds and waves from the northeast as the storm tracks from south to north along the Eastern Seaboard. Sometimes they are referred to as *winter storms* as they usually occur in fall and winter, striking with gale-force winds along any portion of the U.S. East Coast, all the way to the tip of Florida. The famous Ash Wednesday Storm destroyed buildings along the entire coast between Massachusetts and North Florida. The size of the storms, however, is greatest to the north, and they occur more frequently there as well.

Nor'easters are ranked on the Dolan-Davis Scale (named after its authors) from 1 to 5, described as weak, moderate, significant, severe, and extreme based on wave size and duration. Both the Perfect Storm and the Ash Wednesday Storm were extreme (category 5), and grew into this category because

Houses damaged at Camp Ellis, Maine, as a result of the October 1991 Perfect Storm.

Overwash fans on the Outer Banks of North Carolina were the result of a hurricane. —Photo by Andy Coburn

they were very large (the Perfect Storm was smaller, but more intense); they stayed in one area and pounded the beaches for days; and they struck during the highest monthly tides, which meant water levels were higher, allowing big waves to do their damage over a broader area.

The other big storms that East Coast beaches must contend with are hurricanes. As far as beaches are concerned, the most important difference between nor'easters and hurricanes is that hurricanes usually pass over the shoreline quickly, whereas nor'easters tarry awhile and pound beaches for a much longer time. As a result, nor'easters generally produce greater changes in terms of beach erosion (for example, scarping, narrowing, retreating) than hurricanes. In addition, the Perfect Storm, for example, had a very large fetch, or area of open ocean water over which strong winds blow and big waves are generated.

Hurricanes, on the other hand, are more likely to destroy buildings and cause massive overwash of sand onto barrier islands because of their intensity, meaning higher wind velocities, higher waves, and greater storm surge, particularly in the region where the storm makes landfall. Hurricanes are ranked from 1 to 5, with 5 being the greatest, by the Saffir-Simpson Scale (also named after the scale's authors). A storm's rank is based on its central atmospheric pressure, wind velocity, and storm surge height.

All hurricanes are dangerous, but those ranked 3 and higher are very destructive events. The following are examples of some of the more recent storms that caused extensive property loss, flooding, overwash, and dune erosion: Hurricane Fran (North Carolina) in 1996 was a category 3 hurricane; Hugo (South Carolina) in 1989, Felix (North Carolina) in 1995, and Opal (Florida) in 1995 were category 4 hurricanes; and Hurricane Mitch (Honduras) in 1998 and Katrina in 2005 were category 5 hurricanes. Overwash fans are persistent evidence of the high water of such past storms, both recent and ancient. These deposits penetrate beyond the back of the beach into the dunes, maritime forest, and even cross-island into the salt marsh or lagoon.

Having made the generalization about the differences between storm types, it is important to note that in nature differences are never clear-cut. For example, Hurricane Dennis hung around the North Carolina coast for three days in 1999 instead of passing across the shoreline quickly as hurricanes are supposed to. Although it weakened to tropical storm strength before coming ashore, Dennis pounded the coast over multiple tidal cycles and caused widespread overwash along the Outer Banks. In 2004, Hurricane Frances stalled over the Bahamas and lashed the east coast of Florida for three days. Likewise, nor'easters are sometimes more destructive than hurricanes when they linger in the same area, as was the 1962 Ash Wednesday Storm, which destroyed more buildings than the average East Coast hurricane.

Multiple storms can have a cumulative effect on the beach. If two storms hit back to back, the amount of erosion during the second storm may be reduced. Typically, a storm (on an undeveloped beach) flattens the beach and moves sand offshore. As a consequence, when the second storm arrives the energy of its waves are dissipated over a broader beach zone and may move less sand about. On the other hand, if the second storm has a big storm surge, it likely will move more sand ashore by overwash because the first storm already moved sand out of its way. It is important to remember that each beach and each storm is unique and exceptions to generalizations about beach behavior abound. The four hurricanes that struck Florida in 2004 (Charley, Frances, Ivan, and Jeanne) essentially destroyed most of the nourished beaches in their paths and caused much erosion on natural shorelines, leaving areas more vulnerable to future storms.

The surf zone during a storm can be a spectacular geological sight, a natural wonder right up there with massive landslides and erupting volcanoes. The surf zone becomes hundreds of yards wide, and several rows of breaking waves may be seen simultaneously. If the winds causing the waves are some distance away, the waves will be regular and fairly evenly spaced. These are

This lobster trap washed ashore at Wells Beach, Maine, as a result of a nor'easter.

called *swell waves*, or swells. If the winds are right at the shoreline, the surf zone will be confused with irregularly shaped waves, perhaps coming from several directions at once. These are called *sea waves*. Seaward-directed rip currents are often present and may be very strong and dangerous to those foolish enough to enter the surf zone at this time. Of course, surfers converge on the expected landing locations of storms to take advantage of the bigger than normal waves. The only death directly due to Hurricane Dennis was of a middle-aged surfer in Florida.

The beach responds in a variety of ways to a storm depending on the loss or gain of sand, and whether the sand is mounded into bars or spread to flatten the beach. In fact, the same beach will respond differently to different storms. Generally speaking, the immediate poststorm beach is flatter than it was in fair-weather conditions before the storm. Flattening is a protective maneuver in which eroded sand is deposited nearshore, causing the water depth to become shallower. This shoaling allows the energy of the breaking waves to be dissipated over a wider surface. Some beaches, for example, those on Outer Cape Cod (Massachusetts), form offshore bars during storms. These bars also serve as protective barriers that dissipate wave energy as the larger waves trip on the bars and break farther offshore.

For storms of short duration, a beach responds differently depending on whether the storm rolls ashore at low tide or high tide. More erosion will occur during a storm that strikes at high tide, when the beach is narrower and wave run-up may reach the dunes or seawall at the back of the beach. All things being equal, for example, storm intensity, the size of the storm waves will depend on the width of the wave-dampening continental shelf. The wider the shallow shelf, the more wave energy lost due to friction with the bottom. The same storm that crosses the narrow, 35-mile-wide shelf at Cape Hatteras (North Carolina) will produce much larger waves than if it passed along the Georgia coast with its 80-mile-wide offshore shelf. The continental shelf off Miami Beach, Florida, is only 5 miles wide, but storm waves are small because the islands and shallow shelf of the Bahama Banks, 90 miles away, blocks or dampens them. Along the rocky coast of Maine, deep water lies adjacent to cliffed shorelines and allows powerful waves to break onshore, unhindered.

Immediately after a storm, a line of seaweed or other debris will be left high on the beach, well above normal high tide. Sometimes seashells will be more abundant than usual, including some types rarely seen on a particular beach prior to a storm. Rocks, shark teeth, fossil shells from offshore rock outcrops, and coral reef fragments may come ashore along with shipwreck timbers and you name it. If ever there is a time you might find a gold doubloon on a beach (an extremely rare occurrence, of course), it is while beachcombing after a storm. Examples of poststorm finds include a mastodon tooth on Onslow Beach near Camp Lejeune (North Carolina); an Ice Age walrus skull at Cape Henlopen (Delaware); and colonial artifacts, including a cow horn on Brigantine Beach (New Jersey). Artifacts from colonial times have been found on other New Jersey beaches, and Civil War relics on South Carolina Beaches. The flattening of a beach can also expose shipwrecks until sand returns after a storm.

Shoreline retreat occurs primarily as a result of storms, and the greatest changes in the beach profile tend to be storm generated. Because storm frequency is seasonal along the East Coast, beaches may show seasonal differences; for example, a narrower winter beach and a wider summer beach. During winter, larger waves from more frequent storms attack the beach and carry sand offshore, often to form sandbars. This reworking of the sand results in a narrower dry beach and narrower intertidal zone, the area between the high tide and low tide lines. The winter beach is also referred to as a *dissipative beach* because the wider area of nearshore shallow water causes waves to break and dissipate their energy. During summer months, when fewer storms

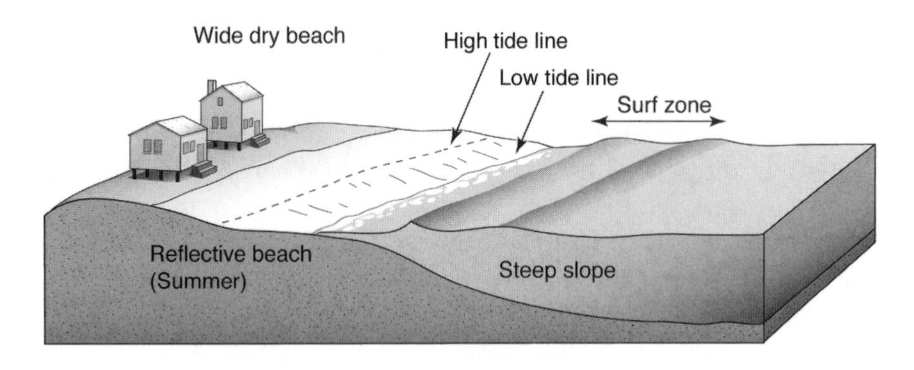


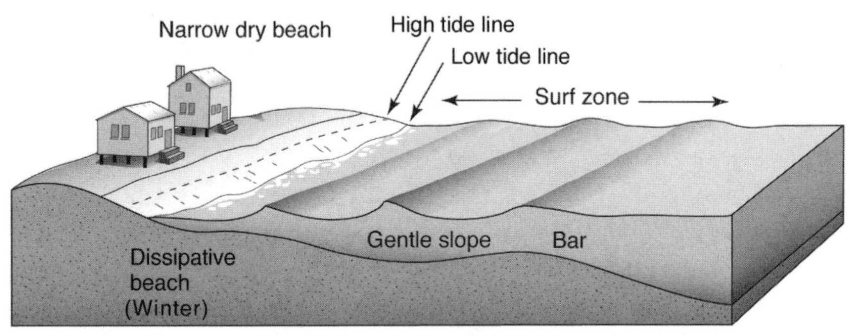


Diagram of a winter beach (dissipative) and a summer beach (reflective). In winter, beach sand is transported offshore, creating offshore bars and a flatter profile. In summer, that sand is returned and the beach grows larger. —Drawing by Charles Pilkey

and smaller waves mean less wave energy, the nearshore sandbar usually migrates onshore and welds onto the beach above water, resulting in a wider beach and a wider intertidal zone with a steeper nearshore slope. This steeper slope results in deeper water near the shore, so there is less dissipation of wave energy due to breaking waves. More wave energy is reflected off the beach face, so the summer beach is also called a *reflective beach*.

Beaches do recover from storms naturally, but usually over time periods ranging from months to years. In the recovery phase, sand is brought ashore and deposited in berms. The recovery can be complete, and in some cases there is more sand on the beach a few months after the storm than there was before the storm. More often, however, the long-term effect of storms is a net loss of beach sand and shoreline retreat, because sand lost to offshore areas in very deep water is lost to the system; also, beach development and engineering structures cause sediment supply to be lost or interrupted.

## Beach Battle Zone: Tides, Waves, and Currents

Along northern New England's shore, storm waves hurl small boulders against solid rock sea cliffs at the back of the beach; the boulders act like battering rams. Along the southeastern barrier islands, hurricane storm surge washes over beaches and into the dunes. In both instances the sea and sand invade the land. In the face of the rising sea level, the Atlantic shoreline is retreating. Indeed, the beach is the front line of a battle zone between the sea, in the form of tides, waves, and currents, and the not-so-solid land's edge.

Even if the global sea level was not rising, the elevation of the sea's surface would not be static. The daily rise and fall of tides creates a zone where the shoreline is exposed to other short-term processes that also cause it to change: storm surge, waves, longshore currents, and rip currents.

### *The Tide*

In a beach visit of even a few hours, you will experience the tide. Tides are the ocean's response to two opposing forces: gravity, the attraction between the earth and the moon, and the centrifugal force of the earth-moon rotational system, which tends to push the earth and moon apart. The ocean bulges

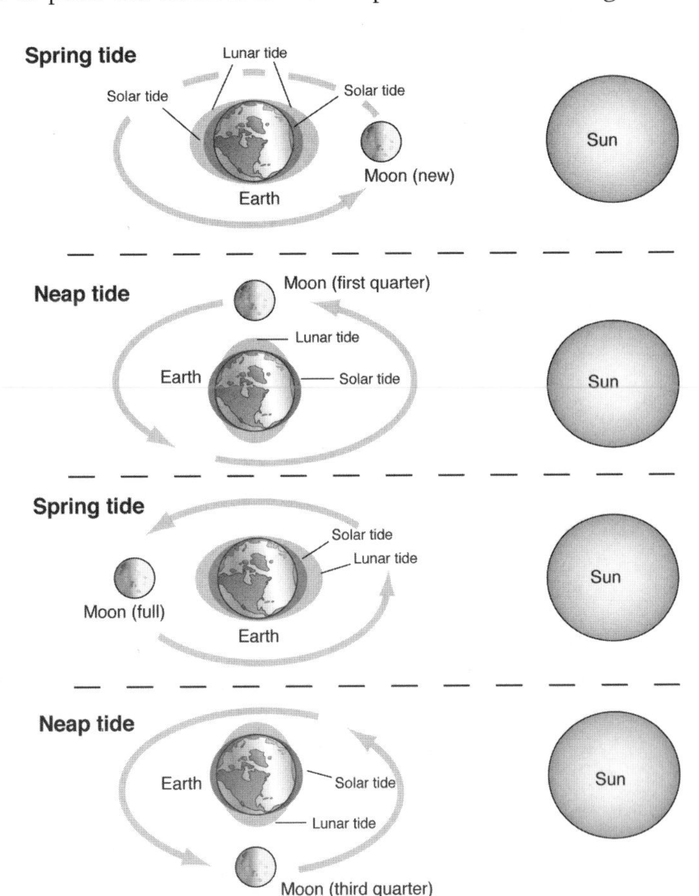


A model of lunar and solar gravitational forces on the ocean, including the conditions necessary for the spring and neap tides. —Drawing by Charles Pilkey

(high tide) on the side of the earth facing the moon because lunar gravity is greater than the centrifugal force, while on the side opposite the moon there is also a bulge (high tide) because lunar gravity is at its minimum and the centrifugal force is greater. Ideally, on the planet at any one time two areas are experiencing high tide, and between them are two areas of low tide.

If the gravitational and centrifigul forces were acting on a continuous, uniform ocean, the tidal cycle might be completed in twenty-four hours, but various complications, such as a delay in response to the tidal forces, continents in the way, and friction between the liquid ocean and solid earth, result in a tidal cycle of approximately twenty-four hours and fifty minutes. A coast that experiences two high tides and two low tides daily is said to have a *semidiurnal tide*, as happens along the U.S. Atlantic Coast. This simplified view of tidal forces is further complicated by the additional gravitational force of the sun on the earth, which can add to or subtract from lunar gravity.

The motions of the earth, moon, and sun largely determine the amplitude (vertical range) of the tide and its periodicity (timing), although the shape of the coastline also has a role. When the earth is aligned with the moon and sun during the monthly full moon and new moon phases, the spring tide occurs. The largest vertical difference between elevations of high and low tide, called *tidal amplitude*, occurs during the spring tide. This is when the coastline will experience its maximum high tides. During the first and third quarters of the moon, the earth, moon, and sun are aligned at a right angle, and the *neap tide* results. Neap tides are characterized by a lower tidal range, and the coastline will experience its minimum high tides during this time of the month.

The tidal range is small in the middle of the ocean, but along the coast the magnitude of the tidal range increases as the broad slope and relatively shallow water of the continental shelf cause tide levels to increase, creating what is known as a *tidal wave*. The spring tidal range along the U.S. East Coast varies from less than 2 feet in parts of Long Island Sound to more than 20 feet along Maine's border with Canada.

The shape of the coast is also important in determining tidal range. For example, the arc of the regional embayment known as the Georgia Bight funnels tidal waters into its central area and creates a large tidal range near Savannah, Georgia. This range decreases to the north and south. The large tidal ranges of 6 to 11 feet along Georgia's coast are in part responsible for the state's vast expanse of productive salt marshes in the intertidal zone, where broad areas of wetland behind the barrier islands are flooded twice daily.

The extreme range of tides in the northern Gulf of Maine is partly related to the funnel shape of the Bay of Fundy, in which tides greater than 40 feet occur. The overall basin shape of the Gulf of Maine also controls the overall large tidal range of that entire region. The Gulf of Maine is just the right size and shape so that as one rising tide floods in and recedes, it is met by the next rising tide. Scientists who measure the movement of ocean water, physical oceanographers, draw an analogy between the tides of the Gulf of Maine and a child on a swing. The resonance of the tides entering and leaving the Gulf of Maine is similar to pushing a child on a swing at exactly the moment the swing returns to its highest position. Push sooner and it is a jerky, short swing. Push at just the right moment over and over and the swing goes higher. This resonance, or coincidence in timing, gives the Gulf of Maine the world's largest tides.

The rise and fall of tides is a great pumping action that creates currents, particularly through the inlets that separate barrier islands, creating ebb tidal and flood tidal deltas. These deltas are great sand reservoirs for barrier island beaches as they evolve.

There is a relationship between tidal range and the types of barrier islands nearby. Reaches of the coast where the tidal range is less than 6 feet are usually characterized by the long, skinny hot dog islands, while drumstick barrier islands occur where the tidal range is between 6 to 13 feet. Tidal ranges greater than 13 feet generate tremendously strong currents, as in the famous Bay of Fundy. These large tides sculpt sand bodies into elongate bars and spits rather than barrier islands that parallel the shore.

### *Storm Surge*

Besides tides, another phenomenon that causes a significant short-term rise in sea level is storm surge, the rise in water level due to low atmospheric pressure and wind and water circulation associated with a storm. The greatest storm surge levels are associated with hurricanes. The sea surface near the center of a storm rises due to the storm's lower atmospheric pressure. The counterclockwise flow of wind around the eye of a storm creates a similar circulation pattern in the sea, causing water to flow around the eye as well as flowing into the center of the circulation gyre. This circulation generates a dome of water, which at sea may only be a foot or two higher than the level of still water, but as this dome of water is forced onshore, the shoaling bottom and irregular coastline cause the surge level to rise higher and higher.

In great hurricanes the storm surge combined with the higher-than-normal storm waves can raise the water level at the coast by as much as 15 to 25 feet, causing flooding for a significant distance inland. The wider the continental shelf, the higher the storm surge. The surge grows in magnitude as it drags bottom along the shallow continental shelf—the wider the shelf, the longer the drag, the greater the build-up. Potentially, the highest storm surges can occur in Georgia where the shelf is the widest (80 miles). If the surge comes in on the high tide, its flooding effect is maximized.

Flooding, storm surge, high tides, especially spring high tides, can all bring areas that are normally above the high-water mark to within range of one of nature's most vigorous agents of erosion: waves.

### *Waves*

Waves are generated by wind blowing across the water's surface. The friction generated as wind blows across the water creates ripples. Given enough time and distance, ripples grow into waves, which can become giant waves under the right conditions. As a wave moves across the ocean surface, the water mass itself doesn't move much; only the wave form moves. Individual water particles remain more or less in place, moving in circular orbits. The crest of a wave is actually the top of this orbital, and the trough of the wave is the bottom of the orbital. Water moves forward as the crest goes by and back as the trough passes.

The size of the waves that come ashore is determined by the fetch, the duration of the wind, and, of course, the intensity of the wind in the area where the waves initially form. Fetch refers to the span of open water over which winds blow while forming waves. The greater the fetch, the bigger the waves. The longer the winds blow and the stronger the winds are to begin with, the bigger the waves.

Wave height is the vertical distance between the crest and the trough of a wave. Wave height is the most commonly reported measure of wave characteristic, and it is a number that people commonly exaggerate, especially when big waves roll ashore during a storm. One way to estimate wave height is to

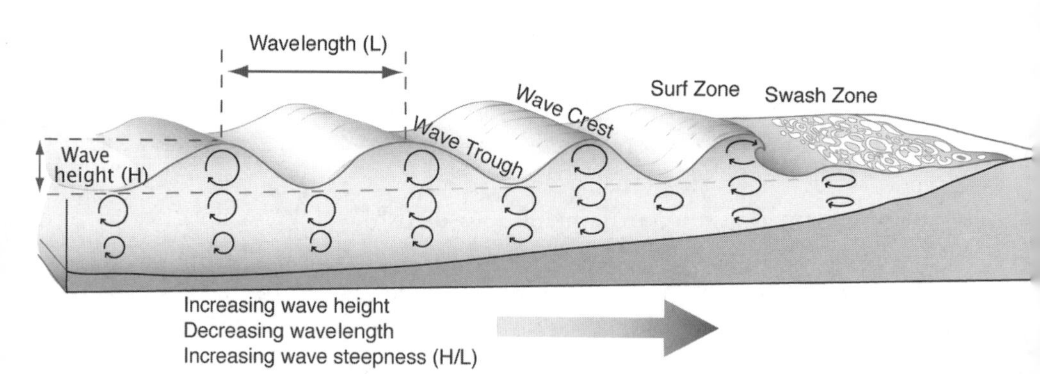


Diagram of a deep-water wave with its crest, trough, wave length, height, steepness, and orbital motion labeled. As a wave moves onshore and encounters the bottom, its orbital motion becomes elliptical and the wave changes shape; it then breaks forming surf and swash. —Drawing by Charles Pilkey

observe troughs and crests as they pass by the pilings of a fishing pier. If there are surfers out on the waves, you can assume that they are 6 feet tall and use them as a scale to estimate wave height. Wavelength is the distance between crests, or from one trough to the next trough. Wave speed is measured by wave period, which is the time it takes two successive crests to pass a given point; it is measured in seconds.

As waves approach the shoreline, wave orbitals begin to encounter bottom, and friction causes the waves to slow down, steepen, and eventually break on shore. In shallow water, the wave orbitals become more egg shaped (elliptical) than round, and the water actually does move forward toward the beach. The orbital motion can be seen quite clearly in the curve of a breaking wave. As a wave rushes onto the beach, most of the energy the wind imparted to the water is transferred to the beach, for example, as surf and swash.

As waves move shoreward and feel the bottom, they slow down and tend to bunch together, that is, their length decreases and their height increases, but their period remains constant. Wave height steadily increases as the water depth grows shallow until the wave literally falls over, or breaks. Usually, after the initial breaking of a wave, smaller, more-closely spaced waves form—waves with shorter lengths, smaller heights, and reduced periods. Each broken, smaller wave finally rushes up the beach in a sheet of water known as the *wave swash*.

A wave coming into shallower water breaks in several fashions, depending upon the size of the wave and the shape of the seafloor in the surf zone. Speaking broadly, waves will break in one of three ways. *Spilling breakers* occur on very gentle beaches. They break when the wave crests oversteepen and spill down the front of the waves. A spilling breaker seems to deteriorate as it approaches shore. A *plunging breaker* is the classic Hawaiian curl-over wave shown in every surfer movie ever made. They occur on steeper beaches when the shoreward face of the wave becomes vertical and the crest then curls off the face of the wave. The plunging water mass expends its energy in a relatively small area. *Surging breakers* break directly on the beach and are immediately transformed into wave swash. These waves do not feel the bottom until they are close to the shore and simply have no time to change shape before crashing into the beach.

Any type of breaking wave can be found on any beach as the nearshore bottom changes shape; for example, when sandbars form and migrate and when other bottom features come and go, or as the waves increase or decrease in size. However, any given beach usually has characteristic wave types depending on the season. For example, plunging waves are common at Nags Head,

Plunging breakers. Osprey with catch for scale. —Photo by Drew Wilson/*The Virginian-Pilot*

North Carolina, and spilling waves are frequently seen along the Georgia coast. Winter beaches, as one might expect, typically have bigger waves due to the greater number of storms offshore, while summer beaches experience gentler waves that reflect calmer conditions.

Storm waves have longer wave periods (10 to 20 seconds) than fair-weather waves (3 to 8 seconds), although the range of these numbers varies a lot from beach to beach. If a storm is being generated right off a beach, the waves striking the shore are likely to be disorganized and confused. Such wave conditions, with multisized waves coming from several directions, are known as *sea*. When the waves are generated from far away, they have the time and distance to sort themselves out and will be more equal in height, shape, and spacing by the time they strike the beach. These waves are the familiar *swell* that we see approaching the shore behind the wave that is currently breaking.

Of course a storm usually produces both sea and swell. For example, during the 1991 Perfect Storm the nearshore waves off New England were described as "sea." At the same time in Virginia and North Carolina, with only light local winds, the huge waves that rolled ashore from the far-away Perfect Storm were evenly spaced swells.

Waves do it all—well, almost all. Waves are by far the most important sources of energy to move sand and change the shape of the beach. As noted,

Spilling breakers.

tides play some role, especially at inlets and where the tidal amplitude is large, as along Georgia and the northern coast of Maine. Of course the wind can slowly but surely move large amounts of beach sand in any direction, but waves, especially storm waves, are the forces that affect beaches the most.

A lot of factors make the waves on any given beach different from those on other beaches. On a regional scale, the nature and frequency of storms in an area is important. In general, the farther north a beach is on the East Coast, the more frequent and the bigger the winter storms. These nor'easters are responsible for much of the change we see in beach shape and beach position over time. Hurricanes, born in the tropics, more frequently strike southern states.

The amount and the nature of wave impact on beaches is often controlled by tides. For example, if a brief storm of a few hours duration strikes a beach at low tide, the impact on the beach will be much different than if the same storm strikes at high tide. The influence of tides on storm wave impact is magnified further when astronomical forces line up and tidal amplitudes are high. When storms arrive at spring tide, all hell can break loose!

### *Currents*

In addition to their role as agents of change, waves also are the progenitors of the currents that are significant transporters of sand. The direction from which waves approach the shore influences how a beach will change. Waves

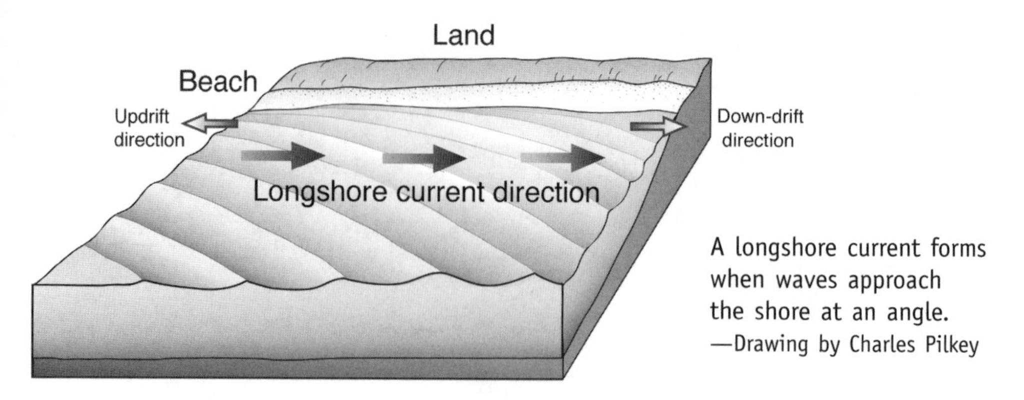


A longshore current forms when waves approach the shore at an angle.
—Drawing by Charles Pilkey

that approach the shore from some angle other than straight on impart some of their energy in a lateral direction as they break, generating longshore currents. As waves roll ashore they slow down when they feel the bottom; if the long crest of the wave is coming in at an angle to the shoreline, one part slows down in shallower water while the rest of the wave moves on at a higher speed. The net result is a bending of the wave, usually making the wave come to the beach more or less parallel to the shoreline but pushing some water parallel to the shore as a current. Such currents carry sand and swimmers alike along the beach. You can see this phenomenon when flying over a beach, or from a vantage point such as a fishing pier or the upper floor of a building.

On most East Coast beaches, currents may carry sand in either direction parallel to the shore depending on the direction and angle the waves approach the beach. Waves coming in from the south, a common summer phenomenon, create currents that carry sand "up coast," or to the north. Waves from the north, which are more common during winter storms, create currents that carry sand "down coast," or to the south. Usually the volume of sand carried in one direction is larger than the volume carried in the other. This difference between the two, the net longshore transport volume, indicates the dominant direction of the longshore current.

Along much of the East Coast, the annual net transport of sand is down coast, or south (west on east-west trending beaches). Major exceptions to this rule occur south of each of the East Coast's great embayments. The longshore currents reverse in these embayments and transport sand from south to north along miles of beach, such as south of the Hudson River mouth (northern New Jersey beaches such as Sandy Hook, Long Branch, and Sea Bright). This same pattern of reversal is also seen at Cape Henlopen and Cape Henry, south of the mouths of Delaware and Chesapeake Bays, respectively.

On Outer Cape Cod (Massachusetts) the currents move sand north and northwest toward Provincetown, and south toward Chatham and Monomoy Island. This regional-scale reversal of longshore currents also occurs because of wave refraction, or wave bending. Off the mouths of the big embayments there are bodies of sand (tidal deltas) dragged out of the bays by tidal currents. These shallow sand bodies cause wave refraction on a large scale and are responsible for longshore current reversal. The same process of sediment transport reversal operates at tidal inlets with a high tidal range to produce drumstick barrier islands.

In New England, numerous rocky headlands and shoals cause longshore currents to vary locally, so there is no regional net directional current. On most pocket beaches, like Sand Beach in Acadia National Park (Maine) there is no net longshore current.

Sand also is transported in the swash zone in a repeated zig-zag fashion. As each wave dissipates into swash on the beach face, the swash runs up the beach at an angle, carrying sand grains short distances up the beach. The backwash travels down the slope of the beach and carries some of those same sand grains back toward the water perpendicular to the beach.

Still another current is generated by waves, one that all too often proves deadly. As a wave moves ashore and breaks, water is pushed ashore. The next wave pushes more water ashore and blocks the backflow of the previous wave. In this way the volume of water builds up in the surf zone, partially creating and being relieved by the longshore current that moves the water laterally. In some instances, enough volume builds up to cause water to flow seaward as a strong current. This phenomenon is known as a *rip current* and is a common threat to swimmers in the nearshore zone. Rip currents usually

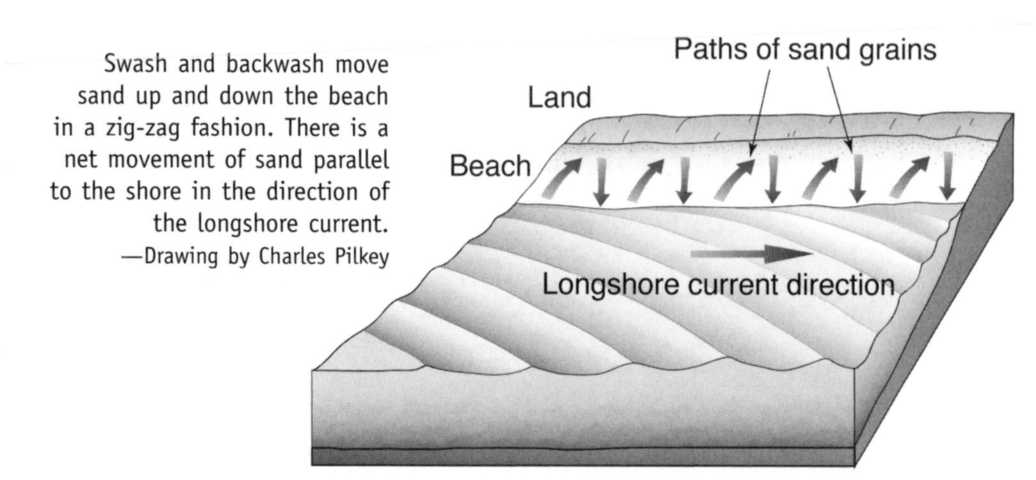


Swash and backwash move sand up and down the beach in a zig-zag fashion. There is a net movement of sand parallel to the shore in the direction of the longshore current.
—Drawing by Charles Pilkey

occur in groups. Sometimes they are evenly spaced and recur in the same location. They are so unpredictable that any warnings regarding rips should be fully heeded. You may be able to spot rip currents; they appear as narrow streams of water, discolored by suspended sediment, moving across the surf zone. If caught in a rip current, you should swim parallel to the shore in order to get out of the current.

Tides, waves, currents, and wind are the agents, the forces, the generators of the many processes that act on coastal sediments to build and shape beaches, dunes, tidal deltas, offshore bars, and other coastal features. And the most common beach material, sand, is discussed in the next chapter.

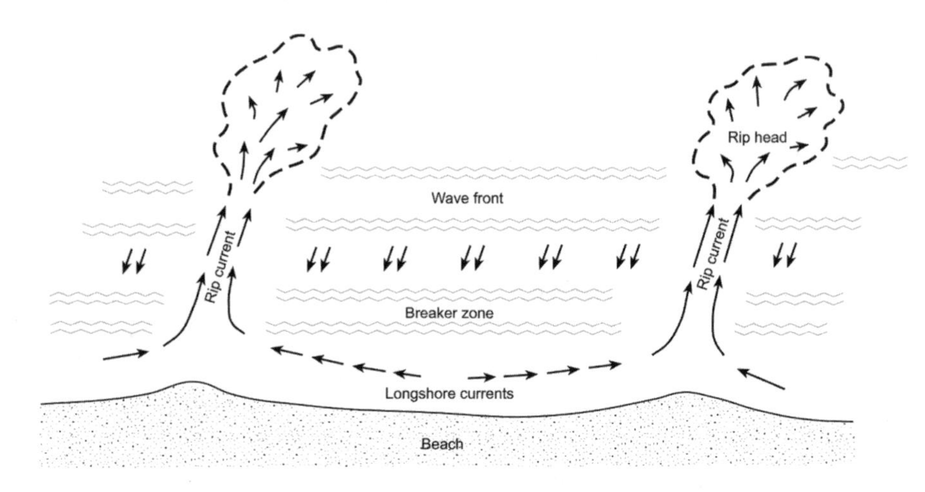


Diagram of rip currents. Successive waves hold more and more water onshore until the longshore current builds up enough force to flow seaward through the incoming waves. —Drawing by Amber Taylor

# Stories in a Grain of Sand

Sand is just the right stuff. The strong ties between sand and recreation in our lives often go back to the sand box, or playing ball on a sand lot, or our introduction to sand on our first trip to a beach. Sand is simply rock and mineral fragments occurring in a specific range of grain sizes. Sand is diverse because the grains vary in mineral composition, shape, sorting, and related properties; and, of course, beaches can be composed of larger particles ranging from pebbles to cobbles to boulders. The overall character of sand on any given beach results in the general properties of that beach—color, texture, porosity, the way the sand grains are packed together—and how the beach responds to tides, waves, and currents.

The following section examines the sources of beach material, how it is modified during its travel to the coast and further changed by beach processes. Again, the geologic setting comes into play in accounting for why beaches of different mineral compositions form, and how this is reflected in a beach's properties.

## As Measureless as Sand on the Seashore: What Sand Grains Tell Us

The most common biblical analogies for amounts too large to measure are the number of stars in the sky and the number of grains of sand on the seashore. Estimating the actual number of sand grains on a beach is an interesting exercise. The Milky Way contains at least 100 billion stars, although we can't see them all with the naked eye. Just how much is a billion? Count the number of sand grains in 1 square centimeter (0.155 square inch) of beach. The result is approximately one hundred grains, with an average diameter of about 1 millimeter. So in 1 square meter of beach (just over 10.76 square feet) there are one million grains of sand on the surface. One thousand square meters of beach surface, an area somewhat larger than a baseball diamond, has one billion grains of sand!

As numerous as sand grains may be, geologists identify, count, and measure sands, sometimes grain by grain, with the purpose of characterizing the sand and ultimately determining its origin. Beach sediment size reflects the sources of the sediment and the energy of the processes that both build and erode beaches. In general, the higher the energy the larger the grains. The mineral composition of the grains tells us something about where the grains came from, and perhaps what the sand grains experienced on their journey to the sea. The properties of the sand grains, such as grain shape and sorting, influence the character of a beach, such as its slope, firmness, water content, or "singing" ability.

## *Beach Sediment: Size, Shape, and Sorting*

Like the biblical authors, most of us equate the beach or seashore with sand. Most people prefer sandy beaches for recreational purposes, but beaches can be made of sediment of any size, including boulders, cobbles, pebbles, sand, and mud. Geologists use a grain-size scale to accurately describe beach sediment, but for our purposes, a pebble is the size of a rock you skip across the water, a cobble is the size of a grapefruit, and boulders can be larger than pickup trucks. For beaches, grain size is the single most obvious and important trait that distinguishes one beach from another.

This shingle beach at Wonderland in Acadia National Park (Maine) illustrates that not all beaches are made of sand. This beach is composed of rock fragments ranging in size from pea gravel to cobbles. Note the roundness of the cobbles that have been sculpted by waves for years.

While it is true that most of the world's beaches are sandy, gravel beaches are also common, particularly along cliffed and glaciated coasts with large, high-energy waves, like those of New England. So-called *shingle beaches* are composed of flattened pebbles and cobbles. Such gravel beaches are common in New England and along parts of the Great Lakes and Pacific shores, but pebbles and pea-sized gravel are found only in localized patches on the more southerly, sandy Atlantic beaches.

The size of sediment on beaches indicates the energy of local waves and the size of the material furnished to the beach. On two beaches with sediment sources of identical grain sizes, the beach with the higher average waves will have larger grains. The bigger waves remove fine sand and transport it offshore or along the shore, leaving behind the coarser, harder-to-move grains. Glacial deposits in New England often supply both gravel and sand to a beach. On these beaches the gravel comes to reside near the base of an eroding bluff of glacial till because waves can't easily move the coarse particles. Sand travels along the beach, far from its source, as seen at Plum Island National Seashore (Massachusetts), or at Jasper Beach in Machiasport, Maine, where glacial sediment is being eroded. Coarse material that does get

The eroding bluff at Highland Light at Cape Cod (Massachusetts) is the natural source of the adjacent beach's sediment. This was taken before the lighthouse was moved back to save it from bluff retreat. —Photo courtesy of Jim Allen of the National Park Service

moved by waves may be reduced in size by the milling effect of breaking waves. Gravel beaches can be quite noisy as the stones are raked and rolled back and forth in the breaking waves and swash, and smaller grains emerge from larger ones.

Gravel exhibits a wide range of forms, or shapes, that reflect the properties of the rocks from which the cobbles and pebbles were derived. The result is an array of shapes, from flat disks to bladed forms, rollers, and equant, nearly spherical stones. For example, if the source rock of a beach's gravel was a slate or other flat, platy, layered rock, the beach stones become platy shingles (disks and blades). No amount of wave washing will ever round these materials, except for their edges and corners, because the rock always breaks along flat planes into disk shapes. Granite, on the other hand, possesses uniform properties in all directions (equant); thus, waves may shape fragments of granite into almost perfect spheres. Other cobbles and pebbles may take on shapes like rolling pins (rollers). Spheres and rollers will roll up and down cobble beaches with the breaking waves, making them the noisiest beaches you are likely to encounter. Quartzites are common rollers.

Waves can segregate material by shape, and this is most visible on gravel beaches. On a beach with a variety of rock types of different shapes, the flat, platy stones (disks and blades) are hurled to the back of the beach by waves and remain there. Rounded stones (rollers and spheres) roll back

Beach materials have been sorted by shape on Jasper Beach in Machiasport, Maine. Waves have hurled the disk-shaped cobbles onto the higher part of the beach, where the person is standing, while the more spherical and spindle-shaped rollers are concentrated lower on the beach to the right.

down the beach slope with the outgoing wave swash and collect at a lower beach elevation.

Sand-sized materials are by far the dominant constituents of East Coast barrier island beaches. The aforementioned relationship between wave height and grain size is neatly demonstrated along the southeast U.S. shore, from Cape Hatteras (North Carolina) to Cape Canaveral (Florida), along the regional reentrant into the coast known as the Georgia Bight. The capes represent areas of high waves, so the coarsest sands are found on Cape Hatteras and Cape Canaveral beaches. Moving from the capes toward the center of the Georgia Bight, sand size becomes gradually finer, reaching the finest average size on the Georgia shore where the waves are smallest. The fundamental reason for this regional variation in wave height and the related change of beach sand size is the width of the continental shelf; it is narrow off the capes and wide off Georgia. Wide shelves dissipate wave energy to produce smaller waves and finer-grained beaches. Narrow shelves provide less protection from larger waves, which winnow finer sediments and produce coarser-grained beaches.

A significant part of the material that is larger than sand on barrier island beaches consists of seashells and their fragments. But even on barrier island beaches, pebbles and cobbles may be found, usually washed in from the inner shelf or brought in as part of artificial beach fill.

At the southernmost beach in New Jersey, at Cape May Point, "Cape May diamonds" (actually quartz crystals) once were collected by tourists. Pebbles of quartz are found on many East Coast beaches. Often these pebbles are flattened, and oval shaped, making good worry stones. The milky white quartz pebbles of Cape Cod (Massachusetts) are reported to sometimes show the property of luminescence or triboluminescence; that is, when two pebbles are struck together or rubbed together in the dark, they glow faintly. Collect a couple of egg-sized quartz pebbles, and when you get home put on safety glasses and go into a dark room. Rub or firmly tap the quartz pebbles against each other and see if they glow.

The larger rock fragments found on the beach may provide clues as to the type of rock that outcrops on the seafloor beyond the surf. For example, pieces of coquina, a shelly limestone, are found on many southern beaches. This light-tan limestone, which formed in Ice Age seas thousands to millions of years ago, is composed of seashell fragments cemented together to form rock. Coquina at Kure Beach, North Carolina, was once mined for road gravel. Outcrops of the rock, which formed during a previous interglacial time, are still on the beach, acting like a natural seawall.

The small white pebbles in the swash zone on this beach at Sandy Hook (New Jersey) are milky quartz. Dime for scale.

In Florida, the rock seen on some beaches is from the Anastasia Formation, an ancient limestone that was used as the building stone for Castillo de San Marcos, a seventeenth-century outpost of the Spanish empire, in St. Augustine. This limestone, also a coquina, outcrops sporadically from St. Augustine to Palm Beach, Florida, and is best seen at Bathtub Reef Park on Stuart Beach, and at The Rocks south of Marineland in Flagler County.

In some localities, beach gravel is derived from ancient river beds. The river gravels were deposited long ago as rivers crossed the continental shelf when sea level was lower. As sea level rose and the barrier islands migrated over these gravel-filled channels, waves moved the river gravels landward and incorporated them into the present beach. In North Carolina, old river-derived patches of pea-sized gravel, including white quartz pebbles, are occasionally found near Nags Head and on Topsail Island. At Sandy Hook in northern New Jersey, similar quartz pebbles are found on the nourished beach, but these pebbles didn't come up naturally from the submerged continental shelf.

Geologists describe the shape of grains in terms of roundness and sphericity. Roundness refers to the angularity of the corners and edges of the grains. Angular or poorly rounded grains exhibit sharp edges and corners in outline, while a well-rounded grain lacks well-defined edges and corners even though it may be somewhat irregular in shape. Such roundness reflects a long history of abrasion, wear and tear from the bumpy ride to the seashore. Sphericity, in contrast, is a measure of how close a grain comes to being a sphere, so spherical grains typically are also well rounded.

Look at a handful of beach sand under a magnifying glass and you will see a variety of shapes and varying degrees of roundness. Some of the smaller angular grains were once larger grains but were broken up during their long transport to the beach. You may also notice that the surface appearance of the individual grains differs; some are dull or look like frosted glass, while others have a high gloss or polish. Frosted grains may reflect a long history of wind abrasion. Grain surfaces may have scratches, pits, and fractures from being mechanically abraded as grains were blown around and rubbed each other; or chemical weathering may have etched or corroded the grain surfaces.

Sorting is another property geologists use to characterize and describe size variation in sediments (for example, beaches may be all sand, or sand and gravel, and so on). Sorting in sands is measured by the range in grain size and provides a way to describe differences between beaches, or between beaches and dunes. If most of the sand grains are nearly the same size, the sand is well sorted. If there is a wide range of sand sizes or if the sand is mixed with mud or gravel, the sediment is poorly sorted. Sorting reflects the range of sediment sizes available and the processes, such as waves, swash, and wind, acting on the sediment.

Dune sand is finer grained and better sorted than the adjacent beach sand, even though dune sand is derived from winds blowing across a wide variety of sediment sizes on the beach. Wind is an excellent sorting agent. It cannot carry the larger and heavier grains transported by waves, but it is effective at moving fine sand. As wind picks up and carries only the finer grains of beach sand, the coarser and heavier grains (for example, gravel, very coarse sand, shells, and shell fragments) are left behind. These coarser grains eventually are concentrated so that they form a lag deposit on the surface of the beach. The lag deposit acts as a protective pavement, or armor, and it inhibits the wind's ability to pick up additional fine-grained sand. Wave action can also armor a beach where it moves sand, leaving behind gravel, cobbles, boulders, and coarse shell fragments.

Mud is composed of clay- and silt-sized sediments, and it is a weathering product derived from rocks in inland drainage basins. One doesn't expect to see mud on beaches because waves keep fine-grained material suspended and move it offshore into deeper water. However, mud and muddy sediments occasionally find their way to the beach, especially after storms, when

Lag deposits of whole shells and broken shell fragments form a pavement that armors the surface of the beach (foreground) as the wind removes the finer, lighter colored sand and deposits it in the dune (background). Camera lens cap for scale.

surf-zone waters carry abundant mud in suspension. Sea-foam bubbles carry clay-sized particles, and after a major episode of foam accumulation, a film of mud can be found on the surface of the back beach after the foam has dissipated. This commonly occurs after storms in Virginia and the Carolinas.

Sediment type, with its many variations in size, shape, sorting, and composition, gives each beach its individual character. Sandy beaches and the near infinity of the number of sand grains may indeed be one of our best models to contemplate "abundance."

## Common Minerals of the Beach: A Mineral Collection in a Handful of Sand

Do you want to hold a handful of gems? Scoop up some beach sand. Depending on where you are on the East Coast, your handful will contain somewhere from three to over thirty minerals, although most of the grains are likely to be silicon dioxide, otherwise known as *quartz*. The sand doesn't appear to be very attractive or very collectible because the grains are so small. Yet if they were larger, they would constitute a good start to an outstanding mineral collection. Some of the minerals are even gemstones, such as garnets, tourmaline, zircon, and epidote.

If you were to sample different beaches along the U.S. Atlantic Coast and identify the minerals at each beach, you would find differences in each mineral collection. Mineralogists term such groups of minerals *mineral suites*. With the exception of shell fragments, the sand grains on beaches are derived mostly from the weathering of bedrock on the continent many miles inland from the beaches. After weathering out of the upland source areas, the grains are transported to the coast by rivers for distances ranging from yards to hundreds of miles. In times past, much of New England's beach sand was brought to the coast by glaciers. Each of these different sources yielded a different mineral suite, so the composition of a beach's sand allows geologists to identify its source rocks in some cases.

### *Mineral Stability*

Different minerals have different degrees of physical and chemical stability. Weaker minerals are altered chemically to clay or abraded to finer sizes when exposed to air and rain (weathering) or while being transported from inland areas to the shoreline. As a consequence, the mineral assemblage that actually makes it to a beach is very different from the original set of minerals that occurred in the source rocks. For example, weathering and abrasion of source material eliminates much of the sand-sized feldspar and other less-resistant minerals. Whereas quartz, which is resistant to abrasion and weathering, becomes more abundant.

Quartz is one of the hardest and most stable common minerals, as far as weathering is concerned, so it becomes more abundant at the beach than it is at the source rock. Because quartz is also one of the most common minerals in rocks, it usually dominates mineral assemblages on beaches. Common dark-colored minerals, such as hornblendes and pyroxenes, which make up igneous and metamorphic rocks, are not as durable as quartz. Weathering readily destroys these minerals, making them much less common in beaches than in the original rocks.

During transportation by rivers, some minerals tend to abrade and break up, especially those that have well-developed cleavage, or planes, along which a mineral tends to fracture. Quartz has a survival advantage in this regard as well, because it does not possess planes predisposed to fracturing. It remains relatively undamaged by transport, even in rushing mountain streams. Pyroxenes and amphiboles from both igneous and metamorphic rocks have well-developed cleavage planes and break up into smaller grains. These minerals also are particularly susceptible to weathering, which leads to clay formation. As a consequence, the abundance of these minerals diminishes during river transport as physical and chemical processes progressively alter them to smaller sizes. While sand and coarser material is concentrated on the

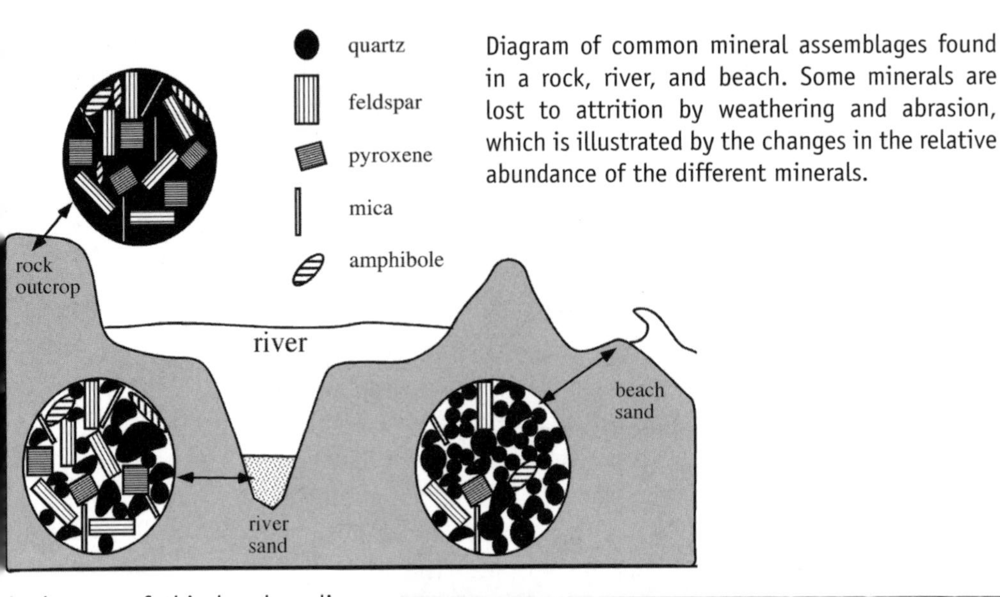


Diagram of common mineral assemblages found in a rock, river, and beach. Some minerals are lost to attrition by weathering and abrasion, which is illustrated by the changes in the relative abundance of the different minerals.

A close-up of this beach sediment shows that most of the sand grains are composed of the mineral quartz. These quartz grains were derived from rocks in the Piedmont and/or Appalachian Mountains, in contrast to the calcium carbonate shell of local origin that will also break down to become part of the beach's sand. The cream-colored quartz grains are stained brownish yellow. The few dark-colored grains are sand-sized rock fragments. —Photo by Drew Wilson/*The Virginian-Pilot*

beach and in nearshore areas, wave energy keeps fine clay-sized sediment in suspension and eventually carries it seaward to settle in deeper water.

Feldspar is another major mineral group that is very common in source rocks, and it is often the second-most common noncarbonate mineral on beaches. *Noncarbonate* refers to the portion of beach sand that contains no seashell fragments, which are composed of calcium carbonate. Geologists use Moh's Hardness Scale to rate the hardness of minerals. On a scale of 1 to 10, with diamond being the hardest mineral (10), feldspar is 6 while quartz is 7. When it is sand sized, feldspar appears almost identical to quartz, and

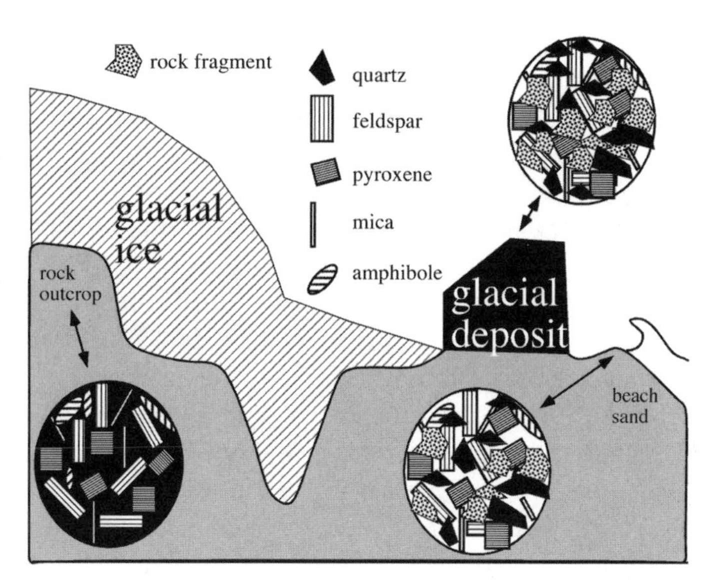


A diagram of mineral assemblages found in a rock, glacial deposit, and beach. Less attrition from weathering and transport occurred because glaciers transported the materials; this is reflected in the relative abundance of both stable and unstable minerals, as well as the presence of sand-sized rock fragments in the beach sand.

sometimes it is difficult to distinguish them even with a magnifying glass. Often feldspar is more common in source rocks than quartz, but its abundance is commonly reversed by the time the grains are rolling around in the surf zone because feldspar is chemically and physically less stable than quartz. On southeastern U.S. beaches, feldspar is usually less than 3 or 4 percent of the noncarbonate fraction of beach sand. That percentage rises in New England beaches, sometimes to more than 10 percent, because feldspar is more abundant in the immediate source rocks and has undergone less abrasion and weathering than the feldspar of southeastern beaches.

Both feldspar and quartz are translucent minerals, which means some light transmits through them, and they are gray to brown in color when viewed on the beach. The color variation is due to iron staining; a thin film of iron oxide can form on grain surfaces, which imparts a light yellowish brown color. If the grains are not stained, they appear to be fragments of clear glass when viewed through a handheld magnifying lens. This presence or absence of staining in quartz-rich beaches results in the sugary white or light tan color of many Atlantic Coast beaches. High percentages of shell material may also give beaches a tannish brown color.

Starting as far south as Connecticut, the mineralogy of northern East Coast beach sand is heavily influenced by the region's glacial history. During the Ice Age, ice plucked much of the sand directly from bedrock sources and then transported it, trapped within the ice, to the coast, where it was deposited. This process of physical removal from the source rocks was very efficient. It reduced the importance of weathering, which is most effective in warm and wet conditions, in the creation of beach sand; glaciation also reduced the importance of physical changes that take place during transportation by running water. Even today, glacier-derived sand is added to beaches as the piles of sediment, left by glaciers in drumlins and moraines, erode along the shoreline. As a consequence, the sand on the more northerly U.S. Atlantic beaches tends to contain a more diverse suite of minerals because the minerals haven't had that much time to weather. In addition, another type of sand grain is common on these beaches: sand-sized rock fragments.

Sand-sized rock fragments are small pieces of fine-grained rocks, such as volcanic rocks (lavas and cemented volcanic ashes), which have not broken down into their component minerals. Individual rock fragments are usually angular and not translucent at all. When a beach has an abundance of rock fragments, it often has a dark color.

As a rule, the gravel, cobbles, and boulders that make up beaches are rock fragments as opposed to being single minerals. On the East Coast, very coarse gravel beaches—not counting shell material—occur exclusively in New England. Some of the large rock particles originated from the erosion of adjacent rock cliffs. Most, however, are derived from old glacial deposits. The rocks can be just about any type because the glaciers that made it to the coast scraped over a huge variety of rocks in Canada and in the northern United States. Careful study of the distribution of these rocks on beaches can sometimes reveal their source areas and the path the glacier followed. Revealing their source areas can also make you rich. Gold! Diamonds!

Mica is another minor, yet sometimes noticeable, component of East Coast beaches. The two main types of mica are biotite, which is black, and muscovite, which is a translucent gray. They come from both igneous (for example, granites) and metamorphic rocks (for example, schists and gneisses). Mica is a flat, shiny mineral that is easily moved by the surf, tending to settle on its flat surface. Mica is a very fragile mineral and is easily broken up into very small fragments, so it is carried out to sea in suspension to settle in deep water. As a result, mica rarely makes up more than a small fraction of 1 percent of beach sand, although it may occasionally exceed 1 percent on New England beaches, because, like feldspar, the parent rocks are exposed in the coastal zone.

Even in very small concentrations, the flat surfaces of mica grains are often responsible for the sparkle of some beaches as sunlight is reflected off the mica flakes. Beautiful sparkly beaches are found on St. Simons and Jekyll Islands (Georgia). The mica flakes, which are less than 1 percent of the beach sand, remain here because of the low-energy waves. In the Carolinas and Florida, mica is usually removed from beach sediment by higher, more energetic waves.

Along with quartz, feldspar, and rock fragments, shell material is a fourth major component of East Coast beach sands. Geologists measure the abundance of seashells in beach sediment by determining the percentage of calcium carbonate, which occurs in the form of the minerals aragonite and calcite. Acid easily dissolves shell material, so geologists determine its abundance by soaking a sample of sand in hydrochloric acid and measuring how much weight is lost after the aragonite and calcite fizz away.

The amount of shell material in a beach is controlled by how much material organisms contribute to the beach versus how much material comes from other sources, diluting the concentration of shell material. This formula is complicated by the fact that many of the shells on beaches are fossils derived from the shoreface, and some shell material dissolves easily in cold northern waters. The abundance of carbonate material ranges from 0 to 100 percent in beaches up and down the East Coast, depending on where you are. Frequently, the sorting action of wind and waves concentrates shells in layers or patches, and these patches often prove to be 100 percent calcium carbonate. Overall, however, the percentage of shell material is highest in South Florida and lowest in Maine. Typical beach sand in South Florida might contain 10 to 25 percent carbonate material, while in Maine 1 percent is closer to the norm.

Calcium carbonate does not just gradually decrease in a northerly direction on East Coast beaches. Cape Hatteras (North Carolina) marks the boundary between high and low concentrations of shells on the beach. North of Cape Hatteras all the way to Maine, typical beach sand, not counting shell pavements that are all shell, contains about 1 percent shell. South of Cape Hatteras, shell fractions are typically 10 percent or higher. Warmer water to the south promotes shell productivity; and there may be less dilution by beach sand because of a smaller sand supply, resulting in higher total shell count.

Having made all these generalizations about the minerals in beach sand, we must recognize the increasing importance of nourished beaches and their impact on beach sand composition. A nourished beach is one in which new sand from some outside source was placed on the beach. An ever-increasing percentage of our beaches are nourished in response to shoreline erosion. South of Cape Canaveral (Florida) close to half of all beaches either have been nourished or plans are afoot to do so. Nearly all of New Jersey's beaches have been nourished. Nourished beaches almost always are different than the natural beaches that preceded them. The most obvious difference is usually a darker color than the natural beach. Whether the mineralogy is different depends on the source of the new sand.

The first nourished beach on Waikiki Beach (Hawaii) is an extreme example of how different a replacement sand can be from the original sand. The sand placed on the beach was imported by ship from Los Angeles! Talk about different. The original Waikiki Beach was almost 100 percent shell and coral fragments, but the new beach was mostly quartz with a few shells. The Miami Beach, Florida, nourishment of 1981 provides a similar example of drastic mineralogy change in beach sand. This change was the reverse of the Hawaiian example. Beaches south of Miami are essentially 100 percent carbonate, but the old Miami Beach was the southernmost beach on the East Coast dominated by quartz sand, about 60 percent. New sand was pumped from offshore areas where the sediment was primarily carbonate sand, shells, and coral fragments, resulting in a new beach with little quartz. The new beach had abundant irregularly shaped particles that packed down into a much firmer beach than the original quartz sand beach.

Sometimes sand for an artificial beach is taken from inland sand dunes some distance from the shoreline, for example, at Virginia Beach. In this case, the heavy minerals and the feldspar content are usually different than the native beach because they have been subjected to thousands of years of weathering on land. More often, however, sand for the beach is dredged from the continental shelf offshore from the beach or from a nearby tidal inlet as part of a channel dredging project.

This section has considered the most common minerals found in beaches, but an even more fascinating aspect of beach sand is the mineralogy of black sand beaches, which we discuss in the next section. The heavy minerals of these beaches are the most colorful, diverse, and interesting minerals and include the gems of the beach!

## Black Sands: Pollution or Provenance?

During World War II, the Japanese sent incendiary bombs attached to balloons toward North America. The balloons were designed to drift from Japan to the forests of the northwest United States and British Columbia, Canada, where, according to the Japanese plan, air currents would no longer hold them aloft and they would crash to the ground, starting forest fires.

The Japanese used beach sand as ballast to adjust the balloons' buoyancy for the cross-ocean trip. Because the ballast included a unique group of minerals in its heavy-mineral fraction, the black sand, geologists were able to

Minerals found in U.S. Atlantic Coast beach and dune sands. Light minerals such as quartz, carbonates, and feldspars usually account for 95 to 99 percent of a typical beach and dune. Heavy minerals, which typically make up from less than 1 percent to less than 5 percent of a beach and dune, occur in greater variety and reflect the composition of the source rocks from which they were derived. Black, green, and purplish sands represent concentrations of heavy minerals called *placer deposits*. In New England, some dark sands owe their color to high percentages of sand-sized rock fragments rather than heavy minerals.

LIGHT MINERALS

- Quartz
- Feldspars (Orthoclase and Plagioclase)
- Carbonate minerals (shell fragments):
  - Aragonite
  - Calcite
- Muscovite and Biotite (Micas)
- Glauconite

HEAVY MINERALS

- Opaque Minerals:
  - Magnetite
  - Ilmenite
  - Leucoxene
  - Pyrite
  - Hematite and Limonite
- Translucent and Transparent Minerals:
  - Olivine
  - Pyroxene
  - Amphibole
  - Garnet
  - Epidote
  - Apatite
  - Staurolite
  - Kyanite
  - Sillimanite
  - Tourmaline
  - Zircon
  - Rutile
  - Monazite
  - Topaz

study the ballast material and pinpoint within a few tens of miles where the balloons were being released. The release site was bombed, and because the American news media had been ordered to not report fires that had resulted from the incendiary balloons, the Japanese assumed the effort wasn't working and abandoned the project.

In order to find the balloon release site, geologists pored over every available Japanese geologic map and report. They compared the minerals in the beach sand ballast to the distribution of surface rocks in Japan that the minerals could have come from. For example, if epidote and sillimanite were common in a sample, then the source rocks were likely metamorphic rocks.

High wave-energy concentrated black sand composed of heavy minerals on the upper part of this beach. Thick layers of heavy-mineral sand were exposed in the face of the trench along the bottom of the photo. The ruler above the trench is 30 centimeters long (about 12 inches).

Zircon or apatite grains indicated igneous rocks. Well-rounded tourmaline and zircon grains could indicate a sedimentary rock source.

Patches of black sand are the most common cause of black coloration on beaches—patches of black that are frequently mistaken for oil pollution. Most often these surface black sands, or the corresponding black layers exposed in shallow trenches dug across beaches, are best developed at the back of the beach near the high tide line, although faint concentrations of black sand are common in the swash zone.

Such black sand consists of heavy minerals. Because of their weight, or more specifically their higher density, the sorting action of both wind and waves has segregated them into layers or patches. Water has a specific gravity of 1, while quartz, the most common component of East Coast beaches, has a specific gravity of 2.7. That is, the mass of an equal volume of quartz is 2.7 times greater than water. Heavy minerals have higher specific gravities ranging from 3 to 5. For example, andalusite has a specific gravity of 3.1, topaz 3.6, corundum 4, magnetite 5.2, and gold 19.3. Because of this difference in the density of individual mineral grains, wind and water—even of low to

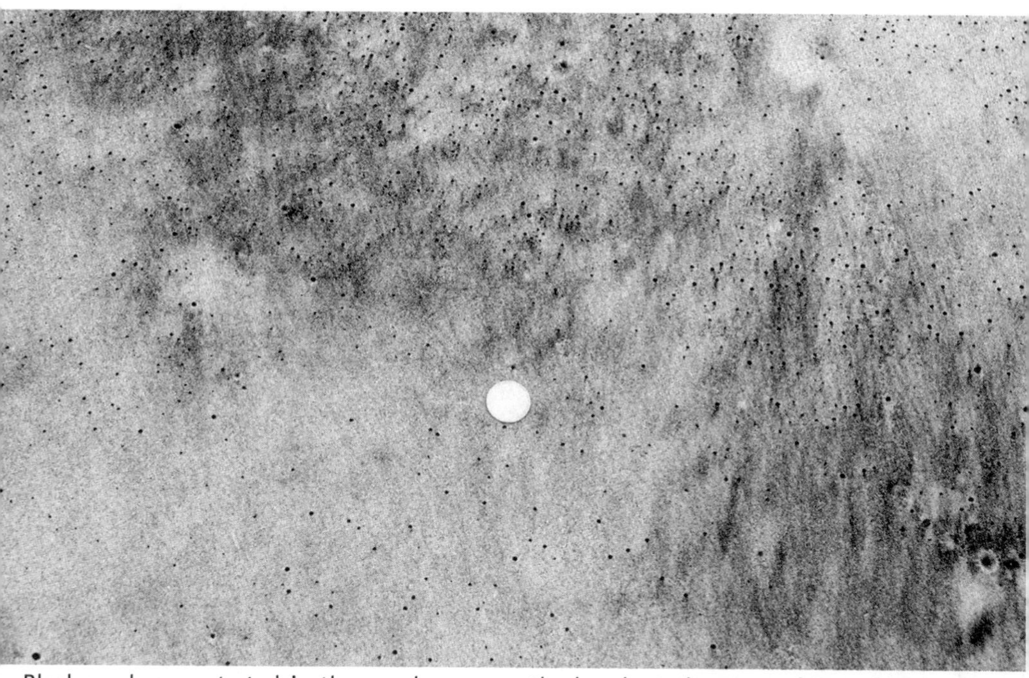

Black sand concentrated in the swash zone on the beach at the town of Emerald Isle on Bogue Banks (North Carolina). Backwash produced the linear pattern. Abundant nail holes are also present. Nickel for scale.

moderate intensity—move light grains, such as quartz. With the less dense quartz grains gone, the more dense minerals are left as dark patches on beaches. Such sorting by density is called *placering*, and after a storm the black placer deposits at the back of a beach may be several inches to more than a foot thick.

You can demonstrate for yourself that black sand is heavier than "normal" beach sand. Carefully scrape up a handful of pure black sand from a dark patch and a handful of light-colored beach sand. The dark sand will feel heavy relative to the other sand.

You can also see evidence of the selective transport of light and heavy minerals in both wave and current ripple marks on beaches and dunes. The heavy minerals, or "heavies," are concentrated in the troughs of ripples, while the light mineral grains, including fine seashell fragments, are selectively transported to the crests of the ripples because of their lower density. These colorful ripples have a striking light and dark striped pattern.

Wind is also an efficient sorting agent and commonly produces heavy-mineral concentrations in dunes. Depending on strength and duration, wind

Wave ripple marks with heavy minerals concentrated in the troughs of the ripples. These ripples have been modified by the changing wave direction so they have flat tops, and streams of sand have been carried off of the ripple crests by the falling tide. The larger objects are snails, grazing mainly in the troughs of the ripples where organic material is deposited. Quarter for scale.

can blow lighter minerals away. A close examination of the stratification in dunes next to the beach often reveals very fine layering made up of heavy mineral layers one or two grains thick.

Taken as a whole, the heavy mineral fraction of the beach and nearby dunes on the East Coast is usually less than 5 percent of the sand. Heavies are much less common in Florida beach sands than in sands from more-northerly states. There are two reasons for this difference. Florida beach and dune sediment is diluted by a larger percentage of shells. In addition, rivers that flow to the Florida coast are very poor in heavies because they drain highly weathered sediments on the coastal plain of the Florida Peninsula. Sand from the Florida coastal plain has resided there for thousands to millions of years, during which time chemical and physical weathering largely removed the heavy minerals.

The low content of heavy minerals in Florida beaches is responsible for differences in the appearance of various bed forms on the beach surface. In fact, the surface features related to the release of air through the beach are much more spectacular in beaches north of Florida because of heavy mineral layering in them. New England beaches generally have the most heavy minerals because most of the glacially derived sediment there has undergone less weathering, hence less loss of heavies.

Heavy minerals are from the same source rocks as the quartz that makes up most beach sand. Many of these minerals, such as those from the hornblende and pyroxene mineral groups, are unstable and have been reduced in volume and number during weathering and transportation. Others, such as zircon and tourmaline, are uncommon minerals to begin with but persist because of their great resistance to weathering and abrasion.

The list of common heavy minerals is laced with gemstones. Zircon, rutile, apatite, garnet, epidote, tourmaline, fluorite, kyanite, topaz, and andalusite, among others, have adorned the hands, wrists, and necks of both men and women for centuries. Many minerals in the heavy mineral fraction are mined from placer deposits in beaches and dunes for various elements. Zircon is mined in modern and ancient beach sands for zirconium (Australia), ilmenite for titanium (north Florida), magnetite for iron (Monterey Bay, California) and cassiterite for tin (Indonesia). Diamonds are mined from beaches on the southwest coast of Africa, and gold in the beach at Nome, Alaska, was once part of the Gold Rush scene.

The black color of heavy-mineral sand in eastern U.S. beaches is due mainly to magnetite, a common iron oxide mineral. Magnetite is highly magnetic and can be readily separated from other minerals with a handheld magnet, providing the sand sample is clean and dry. A simple magnet and a handful of black sand can provide beach amusement for children and adults alike.

The same wind and wave sorting processes that separate the heavy and light minerals can separate heavy minerals from heavier minerals. On Bogue Banks (North Carolina) and other beaches in the vicinity of Cape Lookout (North Carolina), beautiful purple sand often occurs on the margins of black sand patches. These purplish sands are concentrations of garnet. On the Outer Banks of North Carolina, a yellowish green coloration of some heavy mineral patches is often concentrations of epidote. On New England beaches, dark green margins around patches of black sand can be concentrations of hornblende. Sometimes one must be on hands and knees to see the subtle differences in color at the edges of black sand patches. Sometimes, though, they can be spotted from tens of yards away!

The large variety of minerals in the heavy-mineral fraction of most beaches allows geologists to determine the provenance, or source, of beach sand using the same techniques that were used to find the balloon-launching site during World War II. In other words, the heavy minerals in a beach are like fingerprints in the sand. Sometimes such fingerprints point to specific rivers as the source of beach sediments.

From New Jersey to Georgia, the heavy-mineral content of the barrier island beaches indicates that the Piedmont and the Appalachians are the principal sources of sand. The Coastal Plain, which is right next to the coast, is a less important source. On the other hand, Florida beaches, which lie far from the Piedmont, have more sand that was derived from the Coastal Plain.

So why do beach sands derived from the Coastal Plain differ from those that originated in the Appalachians and Piedmont? Mountainous and hilly areas like the Appalachians and the Piedmont erode much faster than the low-lying Coastal Plain, so they yield more sand. This sand has a great variety of minerals because it comes directly from a variety of igneous and metamorphic rocks. The sediments of the Coastal Plain were also eroded from the Piedmont, but these sediments were deposited on the Coastal Plain and remained there for millions of years, forming sedimentary rocks. During this long time of residence, many of the unstable heavy minerals were destroyed by weathering, which reduced the variety of heavy minerals in the rocks. Rivers then eroded these rocks and transported sand from them seaward. The more stable group of heavy minerals that remained after the long period of weathering is typical of sands derived from the Coastal Plain. This distinct heavy-mineral content, the Coastal Plain's fingerprint, allows geologists to readily distinguish Piedmont from Coastal Plain sands, such as those found on Florida's beaches.

From Long Island to Maine, the combination of glacial and river sources complicates the puzzle of beach sand provenance. Because glaciers could have carried rocks and sediment to the region from far away, geologists have a difficult time figuring out the source rocks for northern East Coast beaches.

A sand grain in a beach has had a long and arduous journey to the coast with numerous stops along the way. Some grains make it, but others don't. The heavy minerals may start their journey at the source rocks in abundance, but they end up on the beach as a distinct mineral minority. Still, they are the grains that can best tell us the story of how they got to the sea and where they came from.

### *Other Black Material*

Other common black objects on the beach are blackened oyster shells, mud balls, and tar balls. Mud balls, also called *mud lumps*, are of two origins on open-ocean beaches. Often, compacted mud lumps are introduced during artificial beach filling when dredging pulls up mud from offshore deposits. Mud also can be eroded from offshore deposits or natural outcrops of underlying, compacted marsh mud on some beaches (after a barrier island has migrated over it) to produce mud lumps. Clumps of mud tend to round quickly into mud balls that break into smaller pieces, and then they break down into fine grains that are resuspended in the ocean by waves. Mud is generally too fine to stay in place on the beach, but compact mud balls may persist for some time after they form.

Mud balls are commonly found on beaches after big storms, such as on New Jersey beaches after the 1962 Ash Wednesday Storm. These probably came from submerged marsh mud layers on the continental shelf. Particularly extensive outcrops of mud can be seen at low tide on the beaches off Cape Romain (South Carolina); Ossabaw, Wassaw, St. Catherines, and Jekyll Island (Georgia); on most of the Virginia islands of the Delmarva Peninsula; and sometimes along New England beaches, for example, in Massachusetts at First Encounter Beach (Cape Cod), the Slocum River embayment, and Crane Beach at Castle Neck.

If the mud ball is sufficiently soft, grains of sand, shell fragments, or gravel may become embedded in the surface of the ball as it rolls over the beach. The coating of coarser material armors the mud ball. Armored mud balls are more resistant to abrasion and breakdown and will persist on the beach longer than a plain mud lump.

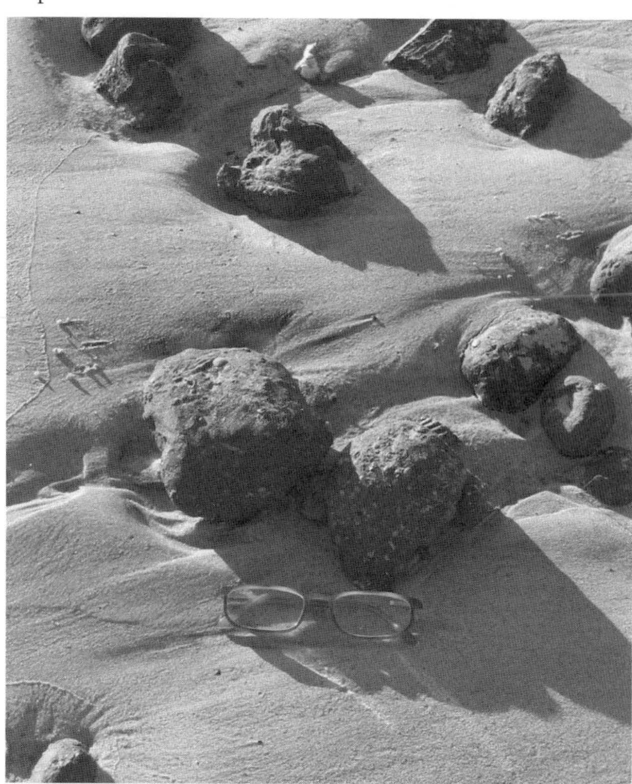

Mud balls from dredge-and-fill sediment pumped onto a North Carolina beach.

Oil pollution can also produce black objects on the beach, most commonly in the form of tar balls. Tar balls are bad news for all concerned. On the East Coast these products of oil spills, small and large, are lumps of semisolid oil that remain after all the lighter components of the spilled oil have evaporated. The tar becomes rounded into something close to a ball, ranging in size from ¼ inch to several inches in diameter, as it rolls about the beach. The oil comes from a variety of sources, including accidental oil spills, deliberate discharges at sea from outlaw vessels, seeps from old sunken wrecks, and even natural oil seeps in a few cases off California.

As spilled oil "matures" with time, it sinks at sea and disappears only to reappear as tar balls pushed ashore by waves. As would be expected, beaches near shipping lanes or harbor entrances (for example, Miami Beach, Florida; Virginia Beach, Virginia; Rehoboth Beach, Delaware; and Sandy Hook in New Jersey) are the beaches most likely to have tar balls. In recent years, however, discharges of oil at sea have been greatly reduced, and the number of tar balls has declined as well. Our impression is that tar balls are much less common on East Coast beaches relative to those bordering the Gulf of Mexico. A decade or two ago, it was common for motels and hotels along southeast Florida, and along New Jersey beaches, to keep rags soaked in kerosene on hand for residents to clean their feet after walking along the beach. If there is tar on a beach, and there is hardly a beach in the world that is totally clean of tar balls, you can rest assured that your children will locate these nasty blebs and return to the car or cottage as tarheels!

Accidents will always happen, such as the June 2003 oil spill from a grounded barge in Buzzards Bay (Massachusetts), in which almost 100,000 gallons of oil were spilled, impacting 94 miles of coastline. Oil and tar balls washed up on beaches as far away as Block Island. Many small spills that make it to the beach are of unknown origin, such as the mystery spill of 1997 that produced tar balls on Assateaque Island (Maryland and Virginia) and beaches to the south. Outlaw vessels still sometimes clean their bilges at sea when no one is watching. During World War II, numerous reports indicate that many beaches were black with oil from vessels sunk by German submarines, especially in 1942 when the U-boats were particularly active and successful in U.S. waters. Tar balls must have abounded on many East Coast beaches for a decade or two after those sinkings.

Armored mud balls buried in a beach deposit at Stone Harbor, New Jersey, after the 1962 Ash Wednesday Storm. Clots of mud were eroded from mud exposures on the beach, then waves rolled them over the beach surface where they picked up a protective coating of sand and coarser material before being buried. These mud balls found their way back onto the beach as the scarp was eroded.

A close-up of one of the armored mud balls from the same exposure. The mud ball is 6 inches long.

# Beach Shape

What do you notice first when you walk onto a beach? You probably get an overall impression based on the geometry of the beach, although you might not use that term. More likely, we would express our impression of the beach in terms of its width, slope, and length, with minor notes on subtle rises and falls of its surface, particularly at the water's edge and out into the surf where sandbars might be apparent. Experienced beachcombers might first notice wrack lines, the flotsam and jetsam of the last high tide or storm that often harbor treasures. And smaller features such as sand ripples may catch our attention, especially at low tide where such features dominate exposed sand flats.

In the following section, we look at the origins of the irregular sand surface many of us know as the beach, introducing common terms used to describe beach shape and defining the aspects of the beach profile. Beach shape and profile reveal the nature, and sometimes the timing, of the events that formed the current configuration of the beach. We examine the character of wrack lines, from man-made objects as large as ship wrecks down to the natural bits of organic detritus that supports beach wildlife. Finally, we review the smaller wave forms known as ripple marks. Ripples occur in a wide array of forms of multiple origin, and they form on different parts of the beach, tidal flats, and associated dunes. All of these structures together reflect a difference of scale in beach processes, from the grander beach profile, expressing overall wave energy; the intermediate scale of the wrack line, reflecting tidal fluctuation; to the more localized processes of breaking waves, swash and backwash, currents in beach troughs, and wind.

If you have the misfortune to be on a groomed beach, many of the interesting natural features will be missing. Beach grooming removes the wrack, robs the beach of its natural forms, and gives the beach a monotonous, level scenery. On the other hand, if you look at the swash zone on a groomed beach, you will see some of the features forming anew.

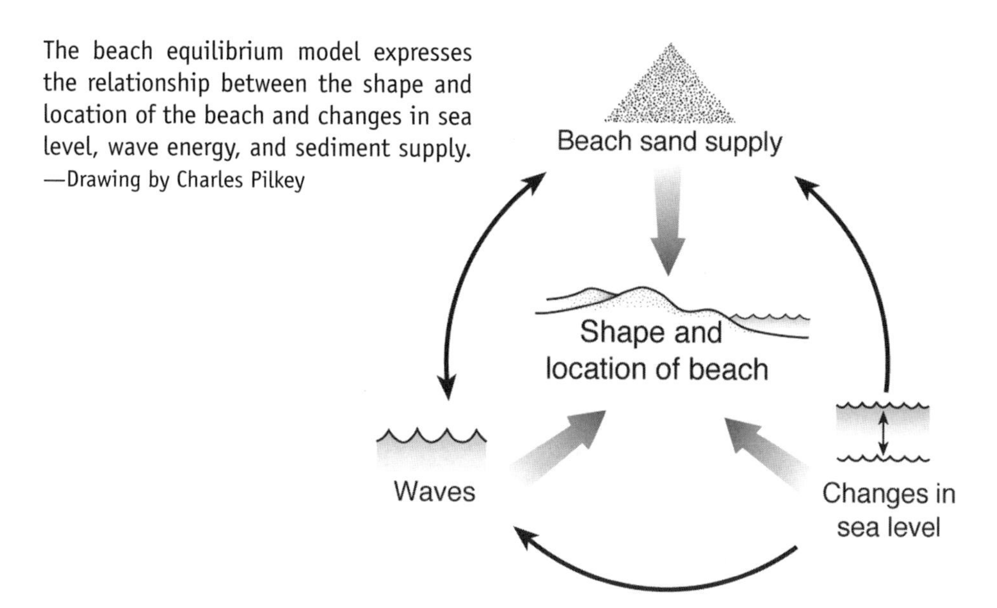


The beach equilibrium model expresses the relationship between the shape and location of the beach and changes in sea level, wave energy, and sediment supply. —Drawing by Charles Pilkey

## The Shape of the Beach: Bars, Berms, and Troughs

Beach shape, and the larger surface features of the beach, for example, the berm, dune, and offshore bars, are expressed as the beach's profile. This is the view of the beach in cross section. As you look down the length of the beach, imagine a line perpendicular to the shoreline and trace out the topography along that line. From the peak of the dune across the beach and to the water's edge, you are looking at the beach profile. Remember that we only see the upper part of the profile above the water line, but a beach's profile continues below water to a distance well offshore, usually to a depth of 30 to 40 feet.

Beach profile is an expression of an equilibrium system; that is, the position and shape of the beach is balanced between sea level rise, wave energy, and sand supply. If one factor changes, the others adjust accordingly. If sea level rises, the beach must move upward and landward, or it drowns. The steepness of the beach slope, including that part of the beach that is below water, changes seasonally with wave energy. The slope is gentler in winter when higher storm-wave energy spreads sands to the intertidal beach and offshore bar, a ridge of underwater sand. The slope steepens in summer when gentler waves move the offshore bar back onshore. Beach width, part of the beach's profile, changes with variations in sediment supply, too; widening when an abundance of sand reaches the shore, or narrowing when the supply is reduced, interrupted, or cut off.

## *Beach Terminology*

An extensive terminology describing the surface features of beaches exists in technical literature. Unfortunately, this terminology is often inconsistent from one textbook to another, or within various governmental and regulatory agencies. As a result, a hodgepodge of terms such as backshore, foreshore, offshore, nearshore, backbeach, forebeach, and so on are applied to the beach with little agreement on definitions among the various users. Of course, nature does not cooperate with those who try to pigeonhole natural features, and the profile of your beach may not match a textbook figure.

We use relatively simple terminology throughout this book. We often note the location of beach features relative to the low and high tide lines, but keep in mind that these can vary widely between spring and neap tides, and with the vagaries of winds and waves. *Lower beach* and *upper beach* are useful terms to describe, roughly and respectively, the lower one-third of the beach surface at low tide and the upper one-third of the beach surface at low tide.

The wet-dry line is also a useful visible reference that shows the extent of beach wetting by the most recent swash. This feature, as variable as it may be, is often mapped from aerial photographs, taken years apart, to determine rates of shoreline erosion. The area between the wet-dry line and the toe of the dune or seawall at the back of the beach is called the *dry beach*. The width of the dry beach is an important beach quality on eroding shorelines, especially those backed by seawalls; on natural beaches the dry beach is wide. Frequently, old seawalls have no dry beach in front of them, even though at low tide there may be a wide, wet beach. A quick glance at the upper beach will reveal whether or not one must pay attention to tide tables in order to be assured a beach to lay on when visiting. At the time of this writing, famous beaches that had long reaches of no dry beach at high tide included parts of Daytona Beach, Florida; St. Simons Island (Georgia); Myrtle Beach, South Carolina; Sandbridge Beach (Virginia); Rockaway Beach (New York); and Wells Beach, Maine.

The most common feature that occurs on the upper beach is the berm. The berm is a terrace or benchlike feature with a steep face toward the sea, and a gentle slope landward to a trough or runnel. The landward and seaward slopes of the berm are separated by the berm crest. Berms can be anywhere on a beach, and beaches can have more than one. Often they migrate up the beach over time, pushed landward by waves. Some beaches never have berms, often for reasons that are not clear. Occasionally, storms form a berm above the normal high tide line. These storm berms may last for a number of months, or until the next storm destroys them. The storm berms that form

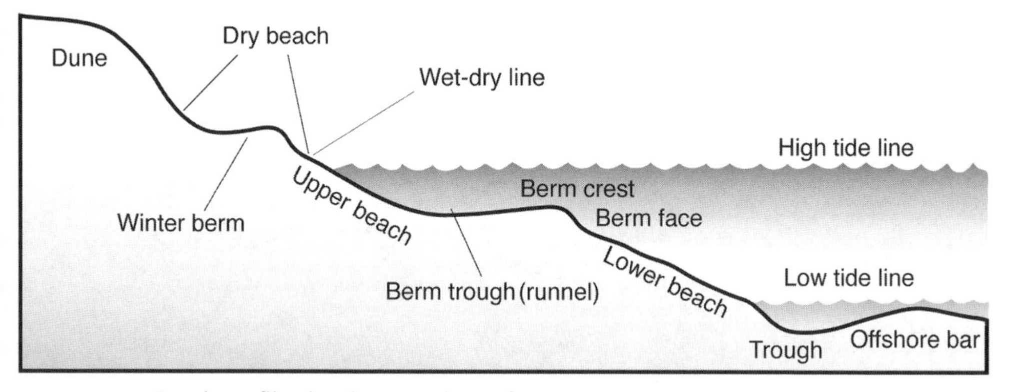


Beach profile showing prominent features. —Drawing by Charles Pilkey

on New England beaches during big nor'easters may persist for years because weaker storm and summer waves can't erode the massive piles of cobbles and pebbles. Usually, the upper berm on the beach is the winter berm, a product of the last big winter storm, and the lower berm, or summer berm, is from the lower-energy waves of summer. (Of course, if you're on the beach in winter, there isn't a summer berm.) Smaller storms may cut storm berms on the upper beach in a steplike fashion down to the summer high tide level.

Beaches with two or three berms appear to be stepped terraces. Multiple berms develop when the berm-formation process repeats itself and older, higher berms are not eroded. Berms can form during quiet weather as gentle waves push offshore sandbars onshore. The sandbar "welds" onto the existing beach, resulting in beach widening and the formation of a new, lower berm below older, higher, more landward berms. This is an example of how increased sand supply (the addition of a sandbar) changes beach width. In contrast, the next big storm may erode sand from the beach, eliminating earlier berms, sometimes reshaping them or moving them farther up the beach.

Surprisingly, after major storms such as hurricanes, the above-water beach is often wider and flatter than the prestorm beach; for example, Long Beach (Oak Island, North Carolina) after Hurricane Hazel in 1954, Shackleford Banks (North Carolina) after Hurricane Fran in 1996, and Hunting Island (South Carolina) after Hurricane Hugo in 1989. In part, this widening occurs because the dunes at the back of the beach are eroded, or because a beach has migrated landward because of overwash. The poststorm profile is also flatter because sand was deposited on the lower beach. Beach flattening during a storm serves a very important purpose. A flatter beach dissipates wave energy; that is, a large wave's energy is spread out over a broad zone

A water-filled trough (runnel) at Ormond Beach, Florida, at low tide. This topographic low, over which the swash is breaking, formed behind a landward migrating sandbar. Note the mirrorlike reflection of the fisherman. This phenomenon occurs when a beach is completely saturated with water.

relative to what would happen if the same wave broke on a steep beach and concentrated all its energy on a narrow zone.

Flattening may be more apparent at Atlantic City, New Jersey, or Daytona Beach, Florida, as opposed to Sandy Hook (New Jersey) or John D. MacArthur Beach State Park (Florida), which are consistently relatively steep beaches. Flattened beaches steepen when an offshore sandbar migrates landward and welds to the beach, causing the beach to widen and its offshore profile to steepen.

The accretion of a sandbar onto a beach can make for an interesting series of observations if you visit the same beach several days in a row. Wade out from the beach to the first bar (be careful going through the deeper trough in front of the bar). Note the distance back to the beach. Repeat the exercise daily at the same place and same stage of the tide. Most likely the crest of the sandbar is getting shallower and closer to the shore, and the trough is getting shallower and narrower. Finally, the bar emerges to form a berm, which may continue to move up the beach if it is pushed by minor storms or spring tides. The trough of the berm will be apparent during low tide when it acts as a drainage channel oriented parallel to the coast and the flow is strong enough to generate ripple marks. The entire process of sandbar migration onshore may take several weeks or longer. On the other hand, some offshore bars tend to remain in place "permanently," as on the Core Banks (North Carolina) near Cape Lookout, perhaps reflecting stronger longshore sediment transport than offshore-onshore sediment exchange.

Some beaches have more than one sandbar paralleling the shore. The sandbar that accretes to the shore is the inner bar, but farther out from the shore is the outer bar. These bars are part of the beach profile. The presence of offshore bars is usually noted by a line of surf where waves first break over them, particularly at low tide. The sudden shoaling of water over the bar crest trips the incoming waves and causes them to break. Off Cape Cod (Massachusetts) in winter, there are frequently three offshore bars marked by three lines of breaking waves. In the beach profile the swash zone is the beach face and is often somewhat steeper than the berm surface.

The shape of offshore sandbars is variable. Bars range from curved and crescent shaped to long and straight, and they can be wide or narrow. The size of a sandbar is important in terms of the volume of sand it can supply to a beach, and the influence it has on nearshore wave and current patterns. For example, sandbars may be linear, but with gaps. These gaps are often the sites of rip currents, where water trapped between the bar and the beach escapes seaward. The alert swimmer can spot these gaps by noting where there are breaks in the white line of breaking waves over the bar crest.

The surf line of breaking waves marks the position of the outer bar along the Georgia coast. The linear inner bar has emerged from the water, attaching itself to the beach in the process of becoming a berm. A trough has formed between the emergent bar and original beach. The trough is flooded at high tide but drains seaward through the gap (center of photo) at low tide. The beach has grown wider.

One thing is for sure: no two beaches are the same and no two beaches respond in the same way to storms or to long periods of quiet weather. Every beach is different. If you want to know what happens on a particular beach, consult a lifeguard or an observant local resident, or watch the beach yourself.

### *Scarps and Cusps*

Two other common features you are likely to see in a beach's profile are scarps and cusps. A scarp is an erosional face, a blufflike feature that waves have cut into the beach or the toe of a dune. Scarps are not particularly common on natural beaches, except at the back of the beach after big storms, because natural beaches are more in equilibrium with prevailing waves than artificial beaches. Scarps are much more common on nourished beaches and range from 1 foot to as much as 10 feet in height. Such scarps on nourished beaches interfere with turtle nesting because the faces of the scarps are too steep for the turtles to climb.

Beach cusps are alternating embayments and beach protuberances that give the beach front a wavy or undulating appearance. Usually, cusps are evenly spaced, often spectacularly so, and anywhere from a few feet to tens of yards across. If you stand on the protuberance between adjacent embayments, you will notice that the wave swash is deflected by the protuberances of the cusps into the embayments. Larger cusps are more likely to form on

Diagram of cusps.
—Drawing by Charles Pilkey

This scarp at the back of Sand Beach in Acadia National Park (Maine) was the result of a storm. Note the wrack atop the upper berm. This scarp is much steeper than the summer berm face.

A scarp that formed on a nourished beach at South Nags Head, North Carolina. The nourishment sand is darker than the native beach sand.

Multiple levels of cusps are rarely seen on sandy beaches, although they are common on gravel beaches. The unusual occurrence of two levels of beach cusps on Edisto Island (South Carolina) in February 2006 is outlined by people standing on the protuberances between individual embayments. The upper row of beach cusps is older and much more widely spaced than the newer row of cusps on the lower beach.

New England's gravel beaches where wave energy is higher. Cusps are often short-lived and change shape and size with changing wave conditions.

Scarps and cusps are but nicks in the larger beach profile, but they illustrate how wave action shapes the surface of the beach. That surface is nature's drawing board, upon which smaller features such as wrack lines and ripple marks are inscribed.

## The Sea's Cutting Edge: Wrack and Ruin

To "draw a line in the sand" is synonymous with a dare or challenge. The shoreline is the sea's "line in the sand," its challenge to the land. Because sea level is rising along the U.S. Atlantic Coast, the land is losing ground, bit by bit, to erosion and inundation. The beach is a scar along the sandy shore that the sea's cutting edge created. In turn, the beach exhibits its own linear features, such as parallel bars, berms, scarps, and the scalloped line of cusps.

Beachcombers are familiar with another prominent line found on beaches. Drift lines, or wrack lines, are the linear piles of flotsam and jetsam that mark the highest water levels of tides and storm waves. The debris, both natural

A wrack line at the back of Big Stone Beach in Delaware Bay (Delaware) is dominated by debris from reeds (*Phragmites* species) and exoskeletons (molts) of horseshoe crabs (*Limulus* species), along with the occasional beer bottle and plastic container.

and that from human activity, is carried by waves and left behind at the uppermost limit of the high-tide or storm-tide swash. Often more than one such line is visible on the beach. The lowest debris line is usually from the last normal high tide, while higher up on the beach another drift line marks the previous high spring-tide level, and still higher at the back of the beach is the drift line of the last storm. If the back of the beach is natural, such as a dune field or forest, you will probably find the remnants of old drift lines well into these areas. Such lines mark the high water of the storm surge in the last big nor'easter or hurricane.

Most of us have a story of some discovery or treasure we recovered from beach debris, such as an unusual piece of driftwood, glass net-floats, whale bones and baleen, a lobster pot, a sand dollar or large shell, and, of course, the usual variety of souvenir baseball hats (our favorite), bottles, sandals, beach balls, and other beach-user trappings. After big storms the wrack is often larger and even more interesting, sometimes containing lawn chairs,

This wrack line from a mid-Atlantic beach consists of an unusual concentration of starfish and wood flotsam, as well as seashells. —Photo by Drew Wilson/*The Virginian-Pilot*

grills, and cottage steps. This human contribution is increasing and comes from both onshore and offshore sources. The offshore component comes from recreational boaters, recreational and commercial fishing vessels, and even the freighters that ply the waters offshore. Runoff from gutters, streets, and storm drains brings material from on land.

Of course, recreational beach users are also a prime source of the litter found in the drift lines. Sometimes the source is obvious, and other times not. Is that exotic beer bottle from an offshore freighter, or was it left behind by a beach-and-beer aficionado?

Plastic—bottles, bags, sheets, ropes—is the most common trash on beaches, probably because we use so much of it and it degrades very slowly. Cigarette butts, though small in size, also are common. Every fall, usually in September, there is an international beach cleanup day, when volunteers collect the trash and refuse on beaches. Often the collected materials are inventoried to identify the sources of pollutants and materials that are detrimental to beaches, dunes, and animals that live on and in beaches.

Not all wrack is trash. A significant portion of the debris line is natural and is usually dominated by various types of seaweed and driftwood from offshore, and frequently by salt marsh straw flushed out of lagoons through nearby inlets. Seeds, shells, and the carcasses of a variety of animals, from birds, small fish, horseshoe crabs, and jellyfish to the occasional porpoise, seal, or large fish are likely to be found in the wrack line. All varieties of seaweed appear in drift lines. Depending on where you are along the Atlantic Seaboard, common forms include sea lettuce (*Ulva* species), which is bright green and lettucelike but feels like waxed paper; mermaid's hair (*Enteromorpha* species), which also is bright green with hairlike stems and attaches to rocks, shells, and pilings; red seaweed (*Gracilaria* species), which

---

In 2003, 10,150 volunteers cleaned 351 beaches across New York state (including lake beaches), and documented over 334,421 pounds of debris, including a wide variety of dangerous items.

| ITEM | NUMBER |
|---|---|
| Bags | 17,097 |
| Balloons | 5,619 |
| Rope | 3,549 |
| Plastic Sheeting/Tarps | 3,380 |
| Fishing Line | 2,202 |
| Strapping Bands | 1,684 |
| Six-Pack Holders | 1,094 |
| Fishing Nets | 557 |
| Syringes | 400 |
| Crab/Lobster/Fish Traps | 350 |
| TOTAL | 35,932 |

---

A sea oats (*Uniola paniculata*) seed waiting to be buried in the wrack line, perhaps to germinate into a sand-trapping plant that will form an embryonic sand dune. The sand grains are quartz. —Photo by Drew Wilson/ *The Virginian-Pilot*

washes ashore in tangled piles; rockweed (*Pelvetia* species), another green algae that lives attached to rocks as the name implies; and kelp (*Laminaria* species), the large, brown algae that make up the so-called kelp forests in the colder waters of New England. Kelp is most common north of Cape Cod (Massachusetts), but it ranges as far south as the eastern end of Long Island (New York); it provides important habitat to animals such as echinoids.

*Sargassum* species, another seaweed particularly common in flotsam south of Cape Hatteras (North Carolina), is a floating algae that lives well out to sea but is carried onshore by storm waves. This plant is particularly interesting because it forms large floating masses that provide a microenvironment for a large variety of other organisms. If you look closely at clumps of *Sargassum* that have washed ashore, you may find little shells attached. When a clump is pushed across the beach by swash, it produces a drag mark in the sand.

People commonly overlook the importance of wrack and drift lines. Perhaps because of the human refuse component, wrack is often viewed as unsanitary and a detriment to the beach. But the natural component of wrack is essential to a healthy beach. It provides the organic detritus that supports other animals that live in and on the beach, the natural biological productivity

A drag mark (lines) produced by a clump of *Sargassum* species seaweed that backwash dragged down the beach. The backwash also caused scour, a process of erosion, around larger objects such as shells. Quarter for scale.

This wrack accumulation of *Spartina* species straw is trapping windblown sand on the north end of Folly Beach (South Carolina), contributing to dune formation.

This groomed beach at Island Beach State Park (New Jersey) has the consistency of coarse powder, is free of wrack and shells, and is as lifeless as a sandbox. The only visible traces of life are the tire tracks and the footprints of a hungry gull. Wrack removal contributes to sand loss on this beach.

of the beach environment. The presence of crabs feeding in the wrack line makes this point. Seeds are a common component of wrack, germinating to produce the pioneer plants that trap sand and form protective dunes at the back of the beach.

The sand-trapping effect of a wrack line is one of its key roles in preserving the natural state of a beach. If you walk along a wrack line, you can see that sand accumulates in the lee of driftwood and similar objects. Sand fills crevices in the debris and sticks to seaweed, straw, and wet surfaces. Ironically, in a time when communities and property owners pay big bucks to put and keep sand on the beach, the daily removal of natural wrack by sweeping and raking also removes sand and the potential to trap more.

### *Shipwrecks: The Ultimate Wrack*

The largest objects to wash up on beaches during storms are shipwrecks. Most of these ships date from the days before modern navigation, and so they are made of wood. Most often, shipwreck timbers reside offshore for many years until the "right" storm comes by and washes the wreck up onto the beach. In 1972, a nearly complete sailing ship came ashore at Core Banks (North Carolina) in Cape Lookout National Seashore during a small storm, complete with some of the rigging and copper-plated cypress planks. The absence of encrusting organisms such as barnacles, borings from other critters, and corrosion suggests that these old wrecks had lain completely

Shipwrecks aren't just large pieces of wrack; they are pieces of history as well. Marine archaeologists glean information from wreck remains and determine the age, origin, and design of the original ship. —Photo by Drew Wilson/*The Virginian-Pilot*

An old shipwreck was exposed beneath a sand dune in Acadia National Park (Maine) in 1986. Sand dunes later covered this wreck.

buried in sand on the seafloor, which preserved them until a storm exhumed the wrecks. After a 2004 storm, the beach was littered with small shipwreck-timber fragments.

If a wrecked boat comes ashore during a bad storm and lands far inland on the beach or sand dunes, it can be buried by subsequent dune deposition. In some places, erosion of coastal dunes by later storms has exposed old wrecks. For example, at Sand Beach in Acadia National Park (Maine), the exposure of such a shipwreck in 1986 was a short-lived affair. After the winter storm season had passed, sand again buried the ship's bones.

## Wave, Wind, and Current Ripple Marks

The 2004 Mars Rover Project identified cross-bedding and ripple marks in martian rocks, allowing geologists to conclude that moving water once covered part of the planet's surface. Similarly, ripple marks on beaches and dunes allow us to identify the fluid processes that have shaped the shoreline in a shorter interim of hours or days.

Like the waves of swell and surf, the surface of dunes, beaches, and the adjacent seafloor are shaped into sand waves of different scales. Waves of sand that form the dune ridges, the ridges and runnels of the beach, and the offshore sandbars are large features. In turn, the surfaces of these landforms are molded into wavy bed forms. A *bed form* is the general term for any small-scale physical feature on the surface of beaches and dunes; technically, a bed form is any deviation from the flat surface. Such features are often preserved in rocks and can be used to interpret ancient environments. *Ripple marks*, the wavy washboardlike surface produced by water and wind currents, including waves, are one of the most common bed forms seen on beaches and dunes. This rippling is analogous to the rippling seen on the sea's surface in a gentle breeze, except it is being molded by waves and currents.

The energy of waves and currents, whether water or wind driven, shapes the sand surface into a variety of dynamic ripple patterns. These ripple marks form as a result of turbulence in the water or air where they contact a sand bed. The turbulence creates unequal forces on individual sand grains, causing them to move by rolling or skipping (saltation). Starting with a flat sand surface, grains begin to move as the water or air speed increases, but not all of the grains move uniformly. Where grains accumulate as microridges, the turbulence increases over the stoss (ocean-facing) side and crest of the microridge, and the ripple grows in size.

Ripples move with water and air currents, too. A current picks up sand grains from the stoss side of a ripple and carries them up to the crest, where they roll down the steeper lee slope or are carried off the crest to settle through the water or air into the trough. In this way, both the individual sand grains and the ripple bed form itself move forward in the direction of the current. The best way to understand ripples is to watch them form, either where water is flowing in a beach trough or where wind is blowing at the back of a dry beach.

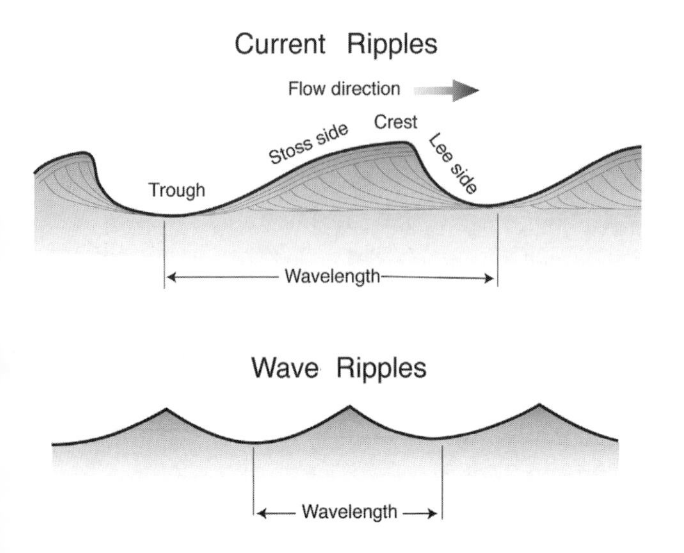


Ripples formed by a current, including shoaling waves, are asymmetrical in cross section. Wave ripples that formed in deeper water tend to be symmetrical due to the oscillatory, back-and-forth motion of water beneath the waves. Very thin cross layers may be seen in cross sections of ripples. The layers dip downward in the direction the current or wave was moving. —Drawing by Charles Pilkey

Asymmetric wave ripples on Jekyll Island (Georgia). Wave ripples have parallel crests, but the ripple crests occasionally split or merge. The steep face of the ripple (left) indicates that the waves were moving to the left. Lens cap for scale.

Wave ripple marks and current ripple marks are the most common bed forms seen on beaches, typically at low tide when extensive sand flats are exposed. Wave ripples are easy to identify because the ripples have parallel crests, giving the beach a washboard appearance. Individual ripple crests may be straight or curved (sinuous). If the water is clear, wave ripples are usually visible just offshore. Actually, wave ripples form in conjunction with the wide variety of currents found in the surf zone, which imparts an asymmetry to the ripples. Because the down-current side of a current ripple is steeper, you can determine the direction waves were traveling when the ripple formed. On tidal flats you may find ripples that formed by currents and waves flowing in more than one direction. Individual ripple crests do not continue on endlessly but may branch into two ripples; conversely, two ripples may merge into one, resulting in a zig-zag pattern.

Close examination of ripple marks usually reveals slight differences in grain size between the ripple crest and trough. Grain distribution on the ripple marks also may vary by grain composition, with the finer, darker, heavy mineral grains concentrated in the ripple troughs. Sometimes lighter organic detritus or mud may also accumulate in the troughs. Ripples with mud in the troughs are called *flaser ripples* and are indicative of deposits formed by tidal currents. The pause in current activity during slack water allows fine debris to settle out to form flaser ripples.

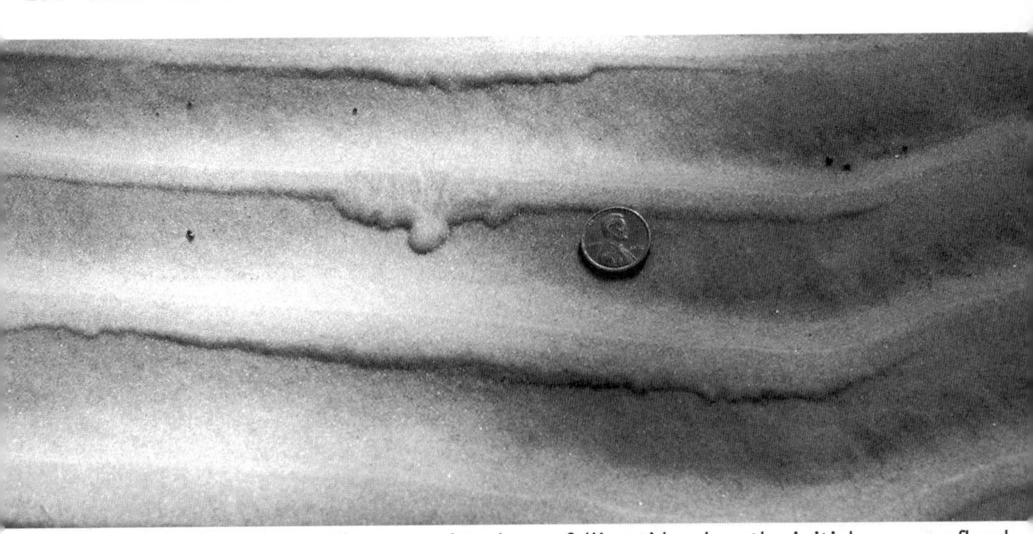

Flat-topped ripple marks usually are produced on a falling tide when the initial wave, or flood-tidal current, ripples are eroded as the direction of flow reverses. Note the irregularity in the ripple to the left of the penny, where some of the sand of the ripple crest has been carried back seaward and deposited in the ripple trough.

Wind ripple marks on this dune surface on a beach at Kill Devil Hills, North Carolina, were preserved by wetting. The surface is pockmarked with raindrop impressions. Quarter for scale.

At the back of the dry beach and in the sand dunes, sand transported by wind forms ripples similar to those formed by waves. These ripples have parallel crests as well. The amplitude, or height, of the wind ripples will be less than ripples formed in water, and they are asymmetric; the steeper face of the ripple occurs in the downwind direction. Concentrations of black (magnetite or ilmenite) or purplish (garnet), heavy-mineral grains in the ripple troughs

A field of ripple marks often reflects several events. First, asymmetric wave ripples formed in shallow water as the waves moved onshore toward the top of the photo (some of the steep faces remain, reflecting the light). The direction of flow reversed (toward the bottom of the photo) on the falling tide, moving the ripples in the opposite direction and eroding off the ripple crests to form flat-topped ripples. Coarser debris accumulated in the troughs. Before being exposed, the ripple crests were breached by the shallow flow and a mix of sand and coarser material was carried through and deposited in the troughs as microdeltas. Finally, a snail moved over the surface, leaving the trace in the trough below and to the left of the lens cap.

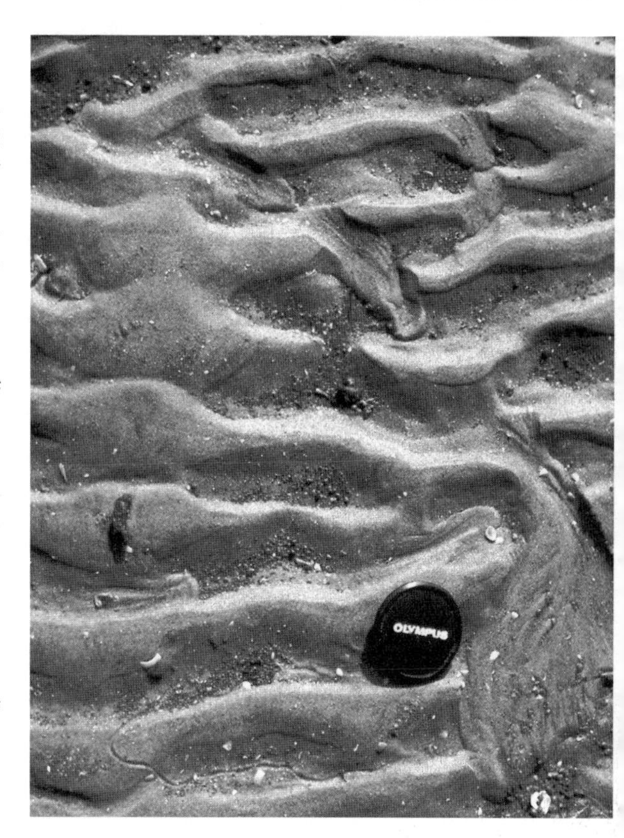


and lighter-colored sand on the crests may result in a striped pattern. The orientation of wind ripples is often different from that of the wave ripples on the same beach because the wind direction differs from the direction of the incoming waves and currents.

On the falling tide, previously formed wave ripples may be partially reworked as currents change or as the water becomes shallower, which modifies the ripple shape or pattern. Flat-topped ripple marks indicate bidirectional flow, in which currents alternately move in opposite directions. For example, with a falling tide, sand of a ripple crest that formed with an incoming tide is carried and deposited on the ocean-facing (stoss) side of the ripple, flattening the crest. Occasionally, a microbreach through a ripple line, almost analogous to overwash on a barrier island, allows sand to be carried through the ripple to the adjacent trough. This process forms a microdelta feature, which looks like a tiny fan or sand lobe in the ripple trough.

If the orientation of the dominant breaking wave or current direction changes, the initial ripple form may be modified or another set of ripples

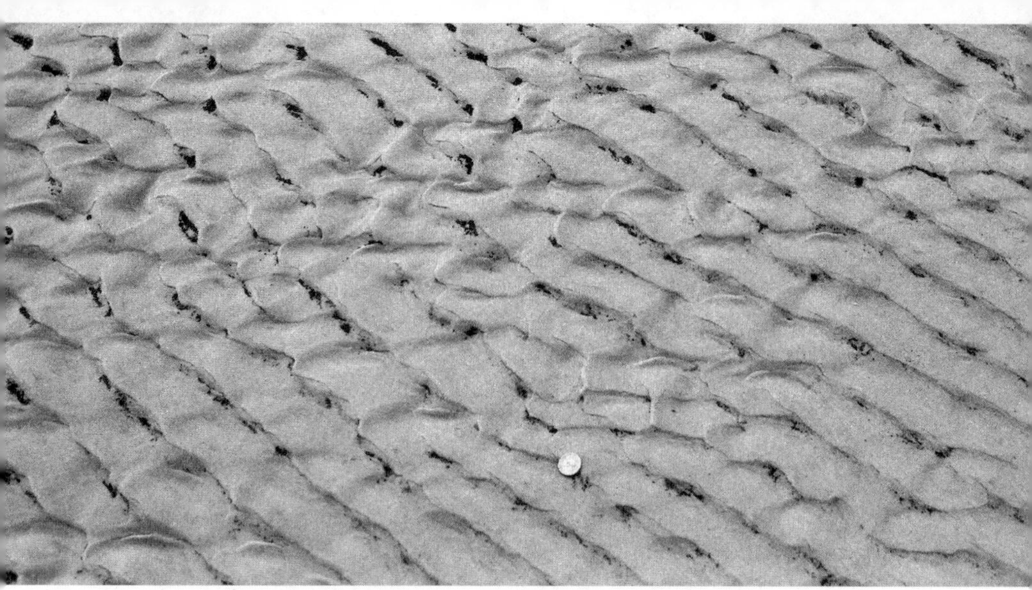

When current or wave direction changes, ripples are modified, or a new set of ripples forms on the previous set. A wide variety of ripple patterns can result. At least three sets of ripple forms are present on this surface. The most obvious ripple set formed due to a current or wave set that flowed from the upper right of the photograph to the lower left. These ripples were modified by a wave set coming from the top toward the bottom/bottom-right, producing the weak pattern of shaded lines running left to right (lee faces) across the upper part of the photograph. The troughs of this second ripple set experienced a right-to-left current, which formed the tongue-shaped (linguoid) ripples (for example, above and to the right of the nickel). The patches of dark material in the ripple troughs is lightweight organic debris. Nickel for scale.

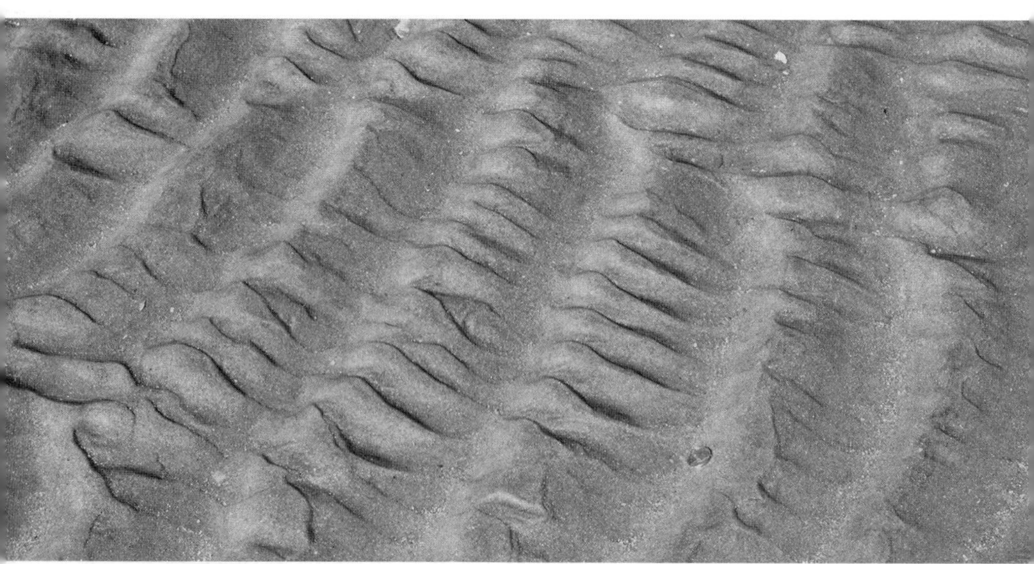

Ladder-back ripples form when one set of ripples is superimposed on an earlier set after current or wave direction changes. This ripple pattern on the low tide beach at Pawleys Island (South Carolina) consists of an earlier set of larger ripples that were moving from right to left, followed by a set of smaller ripples that moved from bottom to top, creating the appearance of rungs on a ladder. Nickel for scale.

may be superimposed on top of an earlier set, producing interference ripples. Such conditions are common on tidal flats where wave directions and currents change rapidly, so that within hours to minutes the current direction can change from a few degrees to 180 degrees (reverse flow). Ladder-back ripples are two sets of ripples oriented at right angles, giving the appearance of ladder rungs. The two sets of ripples indicate that a second wave set, or current, flowed at a 90-degree angle to the initial current. Such ripples are best seen on sand flats at low tide, often near inlets on the surfaces of spits, or on back-island beaches where tidal currents are somewhat gentler and are not strong enough to completely erase the earlier ripple set. Look at the beach at different angles to best see ripples. Shadow and reflected light bring out various ripple patterns.

On a low tide beach, sand flat, or spit, there is often a beach trough, or runnel, in which a current of water is flowing parallel to the beach like a stream; or there may be a channel on the landward side of a spit, which has a seaward-flowing current. In either case, current ripples are likely to form as the current molds the sand surface. Current ripples do not have parallel crests, that is the individual ripples have short crests, are typically arcuate in form, and occur in groups or sequences called *ripple trains*. Shapes are variable and are technically described as linguoid (tongue shaped), cuspate (horn shaped), or lunate (crescent shaped). The discontinuous ripple crests are completely different in appearance from the parallel-crested wave or wind ripples.

Current ripples in a trough, or runnel, between a new berm (left) and the upper beach (right) at Ormond Beach, Florida. Such troughs usually show a variety of ripples.

Watching the flow in a runnel or in a drainage channel where water is draining off the land across the beach is fascinating; it is also instructive in understanding how sand moves and how water current shapes ripples. As water current begins to flow over a flat sand surface, individual sand grains will begin to move along the bed. If you watch for a few minutes, you may observe that more light-colored grains are being moved than dark-colored grains, the heavy minerals. You may also see this selective sorting as coarser material is left behind to form a lag deposit, while the finer sand is moved more rapidly by the flow. As the strength of the current increases, current ripples begin to form, and then gradually move downstream once they are formed. Heavy minerals or organic debris may be concentrated in the ripple troughs. Where the water is flowing fast you may notice a standing waveform, a water wave that remains in the same place. The standing wave develops on the surface of the water and the sand bed as well.

Take a sand shovel and cut a cross section through these ripples to reveal the thin, inclined layers called *cross laminations*. Each lamination is a single layer of sand grains that represents a former steep face of a ripple (lee, or downstream, face). As the ripple migrated, another lamination (a layer one grain thick) accumulated on the ripple face, burying the previous laminations. The end result is a set of steeply inclined (20 to 30 degrees) thin layers. The direction of their downward dip is the direction the current was flowing.

Water flowing in a small channel across this beach reached velocities high enough to generate small standing waves on the water surface (left of foot).

# Can You Read a Beach?

The beach surface is nature's canvas, on which foam, wave swash, wind, air belched from the beach, water flowing out from the beach, or an object carried by some process across the beach create unique patterns, traces, and images. Tracks and trails made by organisms are part of the spectrum of surface features on beaches and dunes, as well, and burrows often appear as internal structures on the faces of our sand-castle moats and trenches. Such burrows crosscut the patterns formed by earlier physical processes. Taken together, the resulting picture from all the holes, bands, burrows, mounds, marks, streaks, shell concentrations, and ledges tells a story that we can indeed read. A beach makes for very interesting reading, much like a detective story or a murder mystery. So when you tire of the book you brought with you, try reading the beach itself.

## Sea Suds: Froth, Foam, and Phantom Marks

We stood at Sandbridge Beach in southern Virginia Beach looking over the seawall at an angry ocean that had been churned into a surface of foam by a small February nor'easter. It was as if a giant box of detergent had been added to the surf, and the excess suds were capping the waves and breaking off in wind-borne blebs. The blebs were blowing over the seawall and accumulating against any barrier that stood in the way. In watching such a display of wave energy and sea suds, one can imagine how such an enigmatic material as sea foam found its way into the creation story of some cultures as the source of life. In Greek mythology, Aphrodite was born of sea foam.

Some miles to the south, in Nags Head, North Carolina, during the same storm, strong swash carried clumps of the foam onto the beach. Balls of foam slid, rolled, and skipped across the beach and accumulated in piles. Some piles were 3 to 4 feet deep against erosional scarps at the toe of a dune. Such accumulated piles of foam lose their white color, taking on a dirty yellowish gray tinge as the bubbles burst and the mass gradually disappears. Those

**Storm waves breaking against the seawall at Sandbridge Beach (Virginia) create airborne blebs of sea foam.**

who walk through the foam end up with a layer of scum on their pants. When the foam dissipates, a very thin layer of gray, sandy mud coats the wrack and back-beach surface.

At the Sandbridge Beach seawall that day, we were observing a common phenomenon associated with high waves and winds that are typical of the stormy winter season in the midlatitudes. Bubbles form readily in seawater as can be seen in any swash zone even in relatively calm weather. Most bubbles that rise to the surface burst and release salt spray into the air, but when they persist and coalesce, sea foam is born. Why bubbles persist is a point of some discussion, but the presence of large amounts of organic matter in the water, particularly the remains of plankton (for example, algae, diatoms, and dinoflagellates), favors the formation of foam.

In heavy surf, the waves' impact on the bottom stirs up a mix of organic matter and tiny particles of clay. Although the seafloor immediately offshore of most East Coast beaches is essentially all sand, small amounts of clay and organic matter are trapped between the grains below the surface. The right storm with the right waves stirs up the nearshore sand and releases the enclosed organic matter and clay particles. Both are the source of the thin mud layer left behind when foam accumulations dissipate on the beach, and both play a role in creating the surface tension that allows individual bubbles, and hence the foam, which is clusters of bubbles, to persist.

Most of us avoid visiting the beach in winter, particularly during storms, so we aren't likely to see spectacular accumulations of foam at the back of

A carpet of foam disappears rapidly after a storm as the bubbles break down. This carpet of cottony foam was originally a couple of feet thick, as can be seen by the mark on the post in the foreground. The foam is grayish due to mud that adhered to the bubbles, and when the foam was gone, a film of mud remained. —Photo by Drew Wilson/*The Virginian-Pilot*

Orrin Pilkey, one of this book's authors, stands in a mass of foam at the back of the beach in Nags Head, North Carolina, after a winter storm. Although this looks like nature's bubble bath, the foam is laden with clay particles and leaves a thin coat of gray mud on the back of the beach—and the author's pants—when the bubbles dissipate.

Foam at the swash edge on Jekyll Island (Georgia). The very faint lineation on the surface of the beach is from individual bubbles pushed across the beach. The foam clump will dissipate and leave a temporary patch of pits on the surface.

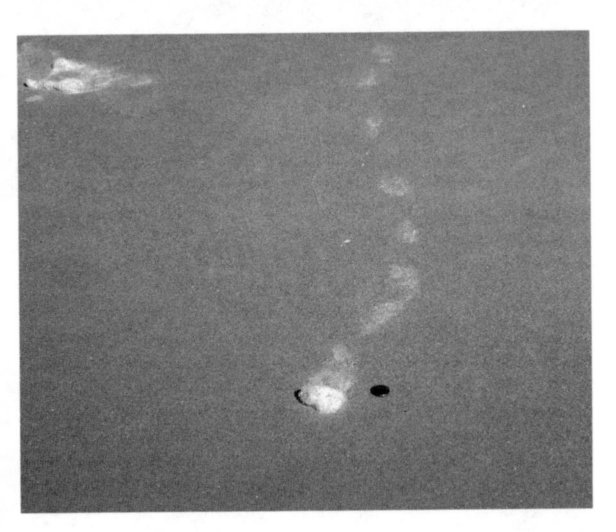

A windblown clump of foam leaves skip marks as it moves across the beach. The pattern will persist after the foam disappears, until the next tide washes it away.

Sediment carried by sea foam is deposited at the edge of dissipating swash, contributing to the formation of a foam swash mark on Pawleys Island (South Carolina). Note the brushstroke pattern between the foam line and the quarter. Such patterns can be the result of passing bubbles or the alignment of sand grains by the swash (lineation).

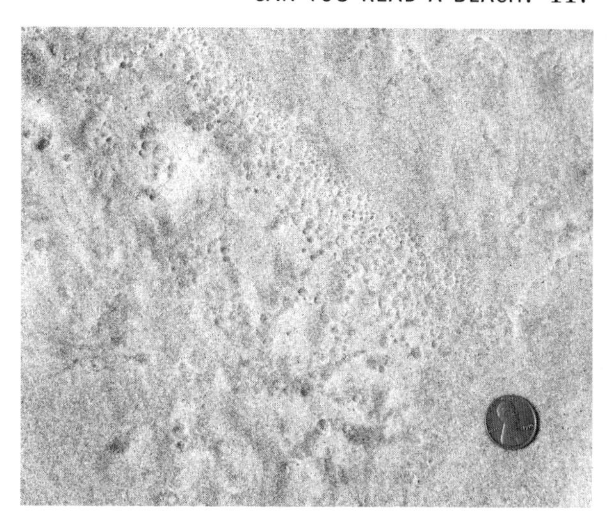

Patches of foam-bubble impressions indicate recent foam on the beach. Bubble pits vary in size and are shallow in comparison to other types of pits and holes. This sand at Hunting Island (South Carolina) is rich in micas (light-colored grains) and heavy minerals (dark grains). Penny for scale.

the beach. But even in fair weather in any season we can see small amounts of foam forming and floating to the beach at the leading edge of the swash. Even a gentle breeze can push small clumps of foam onto and across the beach. On a falling tide, these clumps glide over the surface of wet sand, or they move in a rolling motion; sometimes they even skip across the beach. In doing so, the foam bubbles leave the faintest trace on the surface of the beach, which are mysterious phantom marks to most visitors. Small clusters of bubbles may be pushed continuously across the beach, leaving faint linear trails as well.

Clumps of foam may skip across the beach, and each place where a clump has touched down the beach surface will be a little lighter in color, perhaps because of clay residue. Even a single bubble can move short distances up the beach, leaving a single, continuous bubble trail. Once you know the unique patterns foam migration produces, you can use them to verify recent foam or bubble movement.

The most common foam feature is the foam swash line, a pattern that forms where foam rode wave swash and was stranded as the water soaked into the beach or the wave retreated to the sea. A foam swash line forms a broader line than a swash mark (see next section), up to 1 foot or more in width; a typical swash mark is a line that is only a fraction of an inch wide. The pattern of the foam swash line, produced by bursting bubbles of different sizes, is distinct. When bubbles come to rest, they burst, one by one, forming very shallow pits of varying diameter. You may find the same pattern of pits beyond the foam swash line, where wind transported an isolated patch of foam beyond the swash line. Patches of foam are often found with beach debris, like seaweed, where clusters of foam accumulate.

This interesting scabby, or pitted, pattern sometimes develops at the edge of the wet-dry line. A thin, cohesive wet layer of sand is deposited on dry sand, trapping air bubbles. The bursting air bubbles, plus some curling of the edges of the craters due to surface tension, produces the pattern. Penny for scale.

You aren't likely to confuse pits formed by foam with other similar beach structures. Raindrop impressions are deeper, more regular in size, cover the entire beach surface, and have distinct rims. Splash marks produce impressions that are usually larger than raindrop impressions and much more distinct than bubble impressions. These structures form where waves break against a marsh ledge or some other obstruction on the beach, sending out droplets, or where water drips off driftwood or vegetation overhanging a beach. At the wet-dry line (the high tide line), a less common feature forms where swash carries wet sand over dry sand, giving the surface a pitted appearance. These pits are usually larger than those produced by either bubbles or raindrops.

Where foam has moved over the beach surface, each bubble that touched the sand surface left a very faint track, or line. These parallel lines, discrete bubble impressions, consist of very fine microridges and microhollows and produce a delicate, wavy (crenulated) surface. As the clusters of bubbles move forward unevenly in a series of jerks and stops, each bubble produces a trail (the microhollow) that is approximately its width, separated by microridges. This distinct surface pattern usually occurs on the lower beach during the falling tide or at low tide, and the lines are oriented parallel to wind direction.

Foam structures often are superimposed on other features, such as ripple marks, which were produced by events just prior to foam migration. To look for foam structures, examine the damp surface of a beach on the falling tide, just above the swash line, where foam may be found along the swash mark. The structures you are looking for are very delicate, and you may need to stoop and look at the surface at a low angle to get a shadow effect that will enhance their visibility.

At the back of the beach, or in the wrack line, where foam may have accumulated in quantities sufficient enough to concentrate traces of silt and clay, the surface may have a thin coating of mud. These mud drapes are usually patchy and do not persist very long; they are destroyed by the next high tide, rain, or wind erosion. Before being washed away, the mud layer may shrink (desiccate) and crack to form thin flakes. Where foam is trapped in depressions, mud may be left behind. For example, areas with pronounced ripple marks may have foam bubble patches on the crests and mud drapes in the troughs.

Although foam is not a way for large amounts of sediment to be transported, foam balls can pick up grains of silt, fine sand, and even small shell fragments and carry them to the back-beach area or dune. Along the Irish coast, deposits of silt as thick as 1 foot have been attributed to sea foam deposition. The grains stick to the surface of the foam, producing a foam roll that is analogous to an armored mud ball. The microscopic organic matter the bubbles carry is probably an important food contribution to the microscopic creatures of the beach.

Sea foam and bubbles of the surf make another ephemeral contribution to the beach and dune environments. The saltiness of sea air, one of the invigorating aspects of the seashore experience, is caused by sea spray, which is produced, in part, by bursting bubbles in breaking waves. Sea spray causes salt pruning, the pronounced seaward slope of the top of the vegetation line at the back of beaches and dunes. Salt spray prevents the formation of new leaves and branches on the portions of trees and shrubs that are closest to the beach. Although this gives the impression that the vegetation is leaning landward, it is only the pruning effect of the sea spray. On days when waves are large, the salt spray forms the familiar line of haze that is visible as one looks down the beach.

## To-and-Fro on the Beach: Swash, Backwash, and Wind Structures

Standing at the ocean's edge, one can see a great array of surface characteristics on both the sandy beach and the water. Breaking waves produce a thin sheet of water that runs up the beach, called *swash*; some of the water returns as seaward flow, called *backwash*. The area in which this flow occurs is called the *swash zone*, and the water's to-and-fro constantly modifies the subtle sedimentary bed forms in this zone. Some of the swash may soak into the beach as it flows, depending on whether or not the sand is already saturated with water. Swash effectively sorts different types of grains because the water's

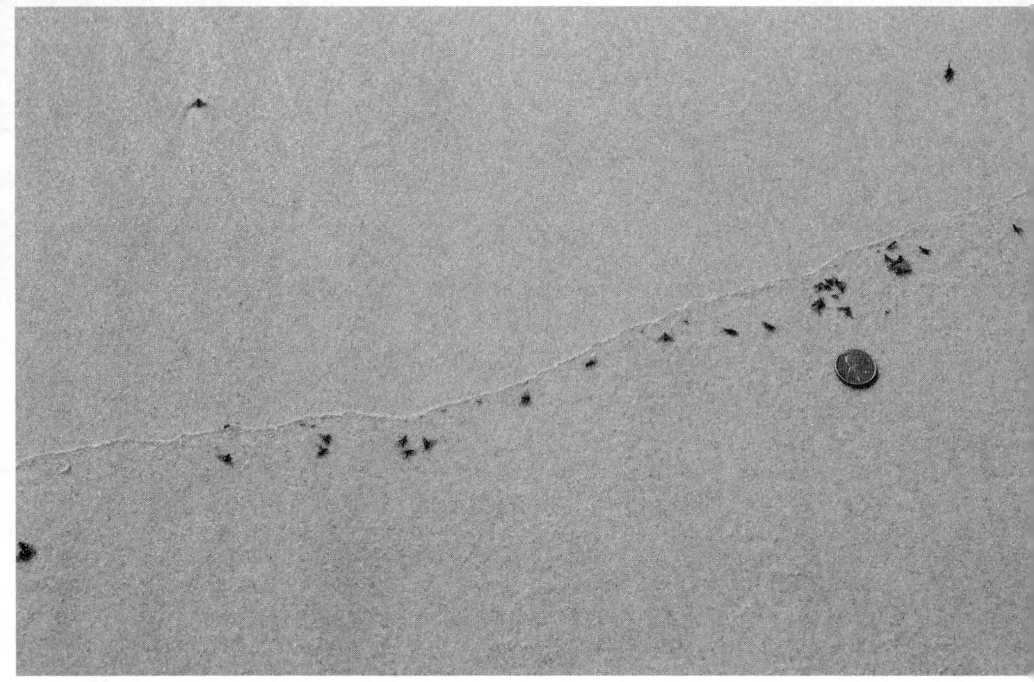

The line marks the edge of the last swash advance, where floating sand grains were deposited to form a microridge (line). Fly carcasses were also concentrated at the edge of the swash.

These swash marks on the beach at Fenwick Island State Park (Delaware) are apparent because of the concentrations of shell fragments.

energy decreases along with its volume and velocity as the swash moves up the beach surface. Thin concentrations of shells or dark heavy minerals commonly appear in the swash zone because the return flow carries away lighter sediments and leaves the coarser, heavier sediments behind.

The leading edge of swash often has bubbles, foam, or tiny bits of floating material; for example, plant fragments, seeds, and bug carcasses. Even sand grains may float on the edge of the swash, held up by surface tension. When the swash dissipates, turns to backwash or soaks in, a distinct line of material—the swash line or swash mark—remains at the point the swash traveled farthest up the beach. Each successive swash produces a new line, which is like a miniature wrack line. The slight difference in grain size between the floating grains and the remaining beach sand is sufficient to produce swash marks that stand out in slight relief and are more visible because of a slight difference in color.

If swash overrides previous swash lines, as with a rising tide, those lines are erased or partially truncated. As the tide falls, a complex series of swash marks are left behind. Take a look at the beach surface downslope from individual swash marks. Often the beach will appear streaked or textured, as if a fine brush or broom had swept over the surface to the swash line. This subtle pattern is produced as the return flow breaks into rivulets and as the gentle swash moves sand grains and orients the grains so their long axes

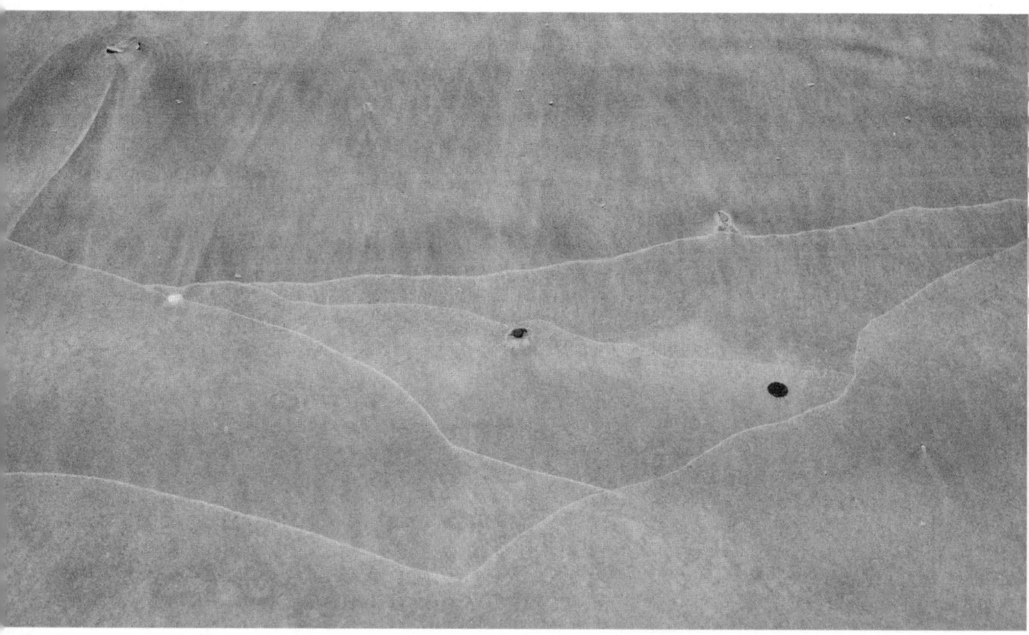

Each new swash line truncates previous swash marks. These marks were made on a falling tide, so some backwash features are still apparent in the upper part of the photo—weak rhomboid ripples and current crescents, which developed from scour around shells. View is looking up the sloping beach face.

The linear orientation of sand grains gives a sense of the direction of backwash over this beach surface. Penny for scale.

are parallel to the lines of flow. The resulting pattern may flare out in a fan shape, with an arcuate swash mark at its edge. Backwash may also produce fine lineation, usually faint streaks due to microgrooves and ridges. The latter may be difficult to distinguish from foam bubble trails.

Where swash just overtops a feature, such as a berm crest, it produces miniature overwash fans that are analogues to the large overwash fans that form when waves wash over a barrier island. The edge of the high tide swash line in this case may have small craterlike structures that are produced when air bubbles trapped below the thin layer of wet sand burst, exposing the underlying dry sand.

Swash and backwash occasionally transport coarser objects such as shell fragments or clots of seaweed across the flat beach, leaving behind drag marks or traces that vary from a simple line to irregular, somewhat mysterious tracks. Sometimes the "tool" that made the mark is at the end of the trace, and sometimes it remains a mystery.

### *Diamond Patterns and Zebra Stripes*

If you walk along the beach during a falling or low tide you might discover a curious and subtle diamondlike or triangular pattern on the beach surface. These V-shaped forms are called *rhomboid ripple marks*, and they form as backwash flows down the beach face. Experts disagree about the details of how these structures form, but rhomboid ripples most commonly occur on steep beaches. The point of the Vs point landward, or upslope. These ripples also form on the landward side of a beach crest or sandbar when waves overwash them. Fields of rhomboid ripples occur on uniform sand surfaces.

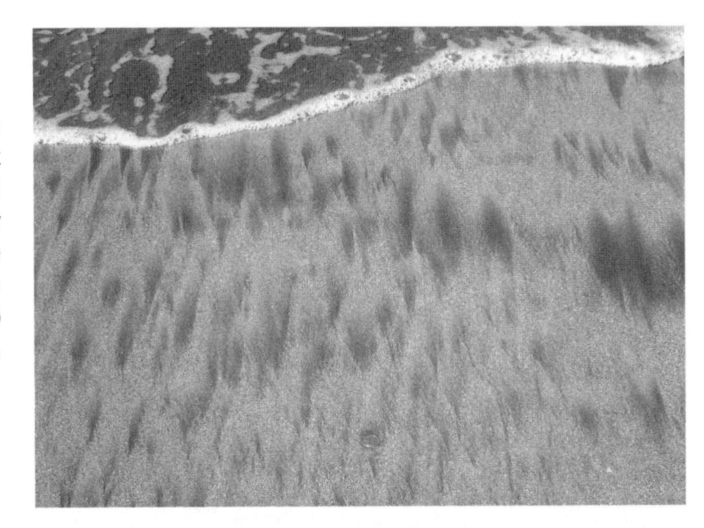

Rhomboid ripple marks have a distinct triangular or V-shaped pattern, and they are common in the intertidal zone on relatively steep beach faces.

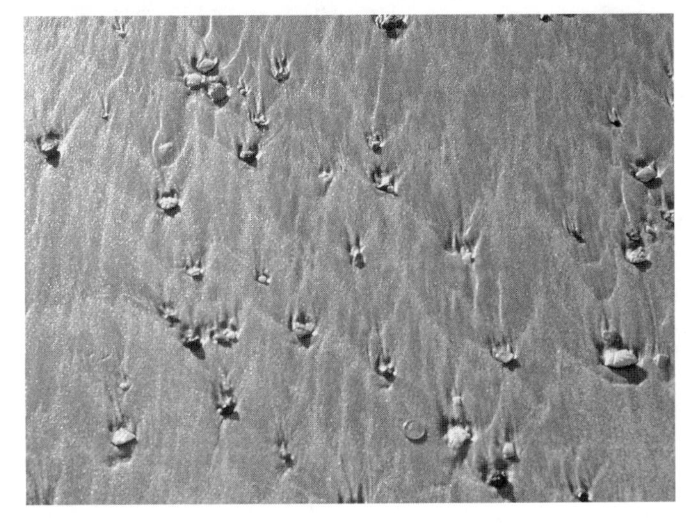

Current crescent marks are produced by backwash scour around objects on the beach, such as these shells at Cherry Grove Beach in North Myrtle Beach, South Carolina. These obstacle marks have a V pattern with the bottom of the V pointing up the beach. Where the obstacles are larger and the backwash stronger, crescents are larger and more deeply scoured.

Similar V patterns, called *current crescents* or *obstacle marks*, are produced by scour that occurs around objects on the beach. A shell, pebble, or clot of seaweed disrupts backwash flow, and the turbulence causes sand to be eroded around the obstruction, producing a V pattern with the wide part of the V facing downstream. This scouring effect is what gives us that sinking feeling as backwash rushes around our feet.

Rapid backwash down the face of a beach on a falling tide creates flow conditions that sedimentologists refer to as *critical*. Watch the backwash and you likely will see riffles forming briefly, parallel to the beach, in the thin layer of water. Little riffles or breaking wavelets form over a bed form known as an *antidune*, so named because its steep face points upstream, just the opposite of other ripples. Antidunes form under critical flow conditions that are dependent on flow velocity and water depth, in which standing waves form and migrate up the beach face in the opposite direction to current flow. These antidunes are typically spaced a foot or more apart. The bed form is short-lived, because the continued flow washes the structure out; however, the footprints of antidunes remain as stripes on the beach as water level falls. The repeated formation of such structures over the span of the falling tide results in the striped pattern that is common on the lower beach. Sometimes dark heavy minerals are concentrated in the troughs between the stripes, enhancing the striped pattern.

Although wind ripples are associated with the back beach and dunes, the wind also forms other intriguing features called *adhesion surfaces* on the damp sand surface of the beach. These bed forms usually develop at the edges of

This riffle line in the backwash flow down the slope of a beach is a breaking wavelet form that briefly creates a bed form known as an *antidune*. Soon after forming, these features are flattened by continued flow, but they leave a tell-tale banded pattern on the lower beach as the tide falls.

A pattern of stripes produced by the erosion of antidunes during the falling tide along the inlet beach at St. Simons Island (Georgia).

arty or pustular adhesion surface forms where wind blows sand grains over the damp beach at Isle Palms (South Carolina). Sometimes called *adhesion ripples*, the surface appears more knobby than pled. The small chimneylike structure in the upper left of the photo is the result of scour around a row wall; it formed before the adhesion structures formed. Penny for scale.

Patches of cohesive wet sand may form as the beach dries out unevenly, resulting in small, biscuit-shaped masses that the wind cannot erode. These features may be the precursors of pedestal structures.

Wind blowing over the damp beach surface on Sea Island (Georgia) produced this strong linear pattern when tiny windrows of sand accumulated in the lee of shell fragments and other coarse particles. The wind blew from the lower right to upper left in this photo. These coarser particles hold up microcolumns as sand is eroded from around them. Note that the pattern gives way to an adhesion surface to the left. Penny for scale.

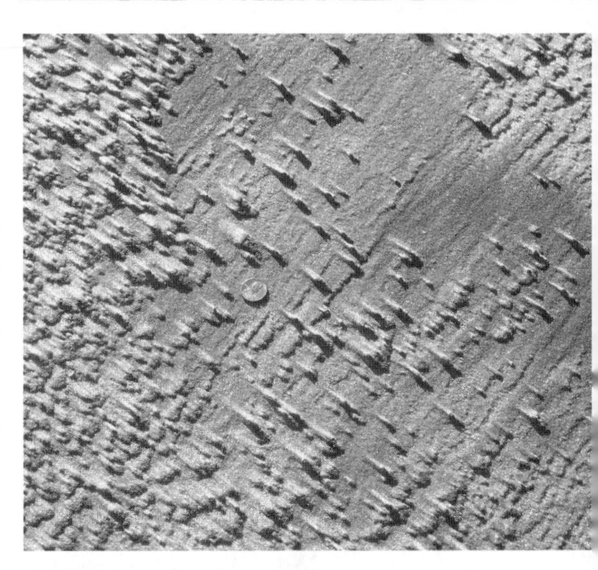

Toadstool-shaped pedestal structures form as the wind removes the surrounding and underlying dry sand from a cohesive biscuit-shaped mass of sand.

runnels and on low tide sand flats where there is a transition from permanently damp sand to sand that is higher up the beach slope and drying. Windblown grains of dry sand carried onto the damp surface stick together. The surface remains damp because water continues to move up to the surface of new sand through capillary action. More and more sand is blown across and sticks to the surface, creating a knobby or lumpy surface. This feature looks more like a field of warts than ripples, but close examination suggests ripple forms with a sense of wind direction.

Similarly, a damp sand surface may favor the formation of sand shadows, which are tiny windrows of sand that accumulate in the lee of shell fragments or other coarse particles on the beach surface. Sand shadows are analogous to the larger sand deposits that form in the wind shadow of vegetation and other obstacles amongst the dunes or at the back of the beach. These features give a strong sense of wind direction.

Another feature that originates from the combination of damp sand and wind is the pedestal structure. Sometimes, what appear to be blisters are actually small, biscuit-shaped masses of cohesive sand; they are held together by moisture in the pores between the sand grains or are partially cemented together by tiny salt crystals that form in the pores as seawater evaporates from the sand. The hard crust of salt-cemented sand grains is called *salcrete*. The surrounding sand is loose and easily blown away by the wind. The cohesive lump protects the underlying sand from blowing away, and gradually a mushroom-shaped pedestal structure is sculpted by the wind. Although not common, a search at the back of an undisturbed beach after a windy day may turn up a patch of these inedible mushrooms.

## Layers in a Ditch: What's Under the Beach?

If the surface of the beach is like a written record, only the most recent page of the story is visible. A vertical cut in the beach is necessary to reveal the pages that tell the story of the beach's past life. Beach scarps or the banks of channels where a creek crosses a beach provide natural exposures of this story, but the pit that kids dig to build a sand castle or the ditch we dig to bury someone in the sand are also good starting points. Just keep your ditch shallow (2 or 3 feet maximum) for safety's sake.

### *Pages of Time: Beach Layers*

Each layer of sand, shells, or rock fragments on the side of the ditch you've dug on the beach was deposited in or very near the surf by some combination

A pit at the back of a beach at Hilton Head Island, South Carolina, showing alternate layers of light-colored sand and dark, heavy-mineral sand. The dark layers represent storms and periods with high wave-energy, when storm waves deposited a high proportion of dark, heavy minerals on the back of the beach. Some of the horizontal layers are cut by animal burrows that filled in with sand.

of waves, swash, backwash, longshore currents, or wind. These processes sort beach sediments by size, shape, and density, which eventually leads to the layering, or stratification, in a beach's structure. Beds are composed of layers, formally called *laminae*. Depending on the process and energy of deposition, these individual layers may range in thickness from the diameter of a sand grain, as small as the size of a period on this page, to more than 1 foot. Sometimes individual beds are not composed of material of uniform size but have coarser materials on the bottom of the bed that become finer higher up. This is called a *graded bed* and indicates that wave or current energy decreased as sediment was deposited.

A bed is a distinct layer that appears to be different from the sedimentary layers above and below it. The difference may be the result of changes in grain size, color, or the nature of the bed's internal laminations. For example, storm waves might concentrate a black, heavy-mineral deposit at the back of a beach on top of lighter-colored beach sand. After the storm, wind may deposit lighter-colored rippled sand on top of the black sand. A trench dug through the back of the beach would reveal these three distinct beds: the lower beach sand, which might contain parallel laminae, the black heavy-mineral sand, and the upper light-colored sand bed that would have internal cross laminae. The top and bottom of a bed, or bedding surfaces, usually result from a change in depositional process. For example, the top and bottom of

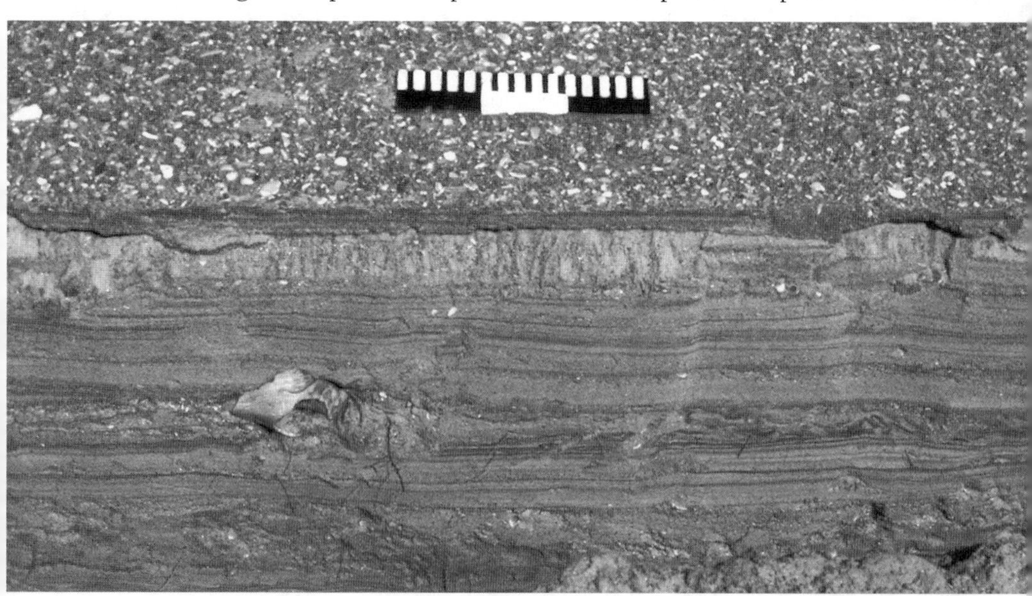

Cross section of a beach at Cape Lookout National Seashore (North Carolina) showing horizontal layers and some inclined layers called *cross laminae* in the dark center layer. The dark layers consist of sand grains of dark rock fragments, while the black layers are rich in heavy minerals. The surface of the beach is a shell lag deposit, and when buried it will be a light-colored layer. Note the buried shell (lower left). Scale is in centimeters.

the black sand bed mark the beginning and end of storm conditions that concentrated heavy minerals.

The color of a layer depends on its composition. Heavy minerals produce black layers, which are sometimes particularly abundant and thick at the upper part of the beach. Along with heavy minerals, dark sand-sized rock fragments may form distinct colored layers in New England beaches. Concentrations of shell fragments produce coarse layers of a variety of colors, usually brownish in southeastern U.S. beaches. Layers of quartz sand without shell fragments will usually be a light gray to grayish brown color.

Layers made up of coarse shell hash (shell gravel) present an enigma. They could have formed during either high-energy or low-energy conditions. For example, energetic waves capable of moving large shells about can deposit a shell layer. Alternatively, the shells may have been concentrated into a distinct layer as wind or wave swash gradually removed, grain by grain, the matrix of finer sand grains that once surrounded the shells, leaving behind a shell layer called *shell lag*. This can then be covered by finer sands and become part of the beach's structure.

Individual layers preserved in the beach may represent an event of a few seconds, as in a thin layer deposited by wave swash or a gust of wind; or they may represent an event of hours, as in the inches- to foot-thick layers of black heavy-mineral sand concentrated on the upper beach during a storm. Or a layer may have formed over a time frame of days, as in the case of shell lag.

The slope of an individual lamina (thin layer) in cross section, of course, represents the slope of the beach surface at the time the layer formed. Most layers slope gently seaward, but layers of a berm often slope gently in a landward direction. Thin layers are wavy where the beach was rippled. If you carefully cut a face across a ripple mark, you can see that its thin layers, or cross laminae, were deposited parallel to the steep face of the ripple. These ripples can become entombed in a beach's structure, and the cross laminae of the ripple's steep face are inclined differently than the overall layering of the beach.

Similarly, cross-sectional exposures in dunes reveal that they have inclined beds, called *cross-beds*, that are much thicker than the cross laminae of ripples. Cross-beds accumulated on the lee, or slip, face of the dune and are typically inclined at an angle of about 30 degrees, usually in a landward direction. If you find a cut through sand dunes, most likely you will see sets of cross-beds that cut into earlier cross-bedding, and the inclinations of the different beds will vary. In reality, dunes don't always resemble textbook illustrations. You will find dune faces in all orientations with beds ranging from perfectly flat to 30 degrees.

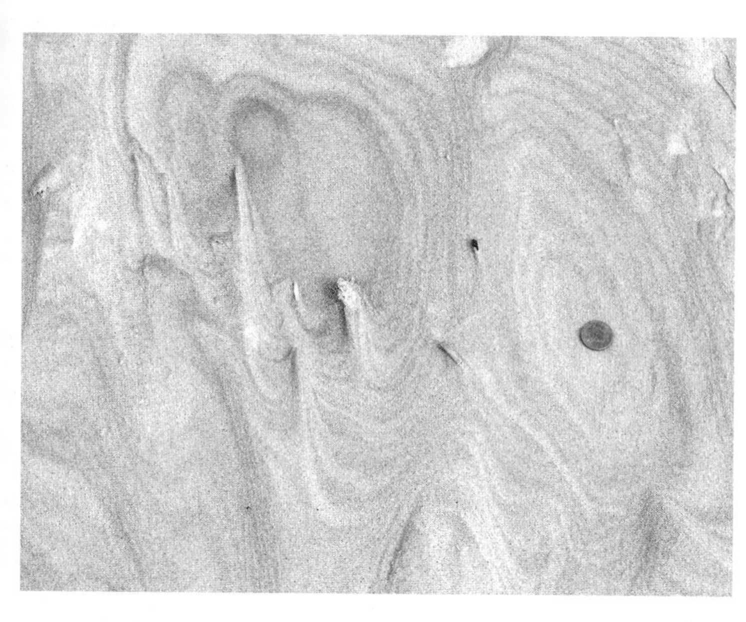

The contour pattern in th surface of this beach at Hilton Head Island, South Carolina, was caused by the erosional exposure of laminae of different colors. Penny for scale.

Sometimes you can see evidence of a beach's different layers without digging a ditch or finding a scarp or drainage channel. This is especially true when the beach is underlain with alternating layers of light-colored sand and dark sand rich in heavy minerals. Where the wind differentially erodes the surface, or where water seeps from the beach, the microtopography cuts through these thin alternate layers to produce patterns that look much like the contour patterns of land surfaces seen from an aerial view.

### *Bioturbation*

The original layering of a beach or dune is always disrupted, to some degree, by burrowing organisms. This disruption is called *bioturbation*. The animals include clams, especially the coquina and razor clams; worms; crabs, such as mole crabs; and ghost shrimp. Different organisms inhabit different parts of the intertidal zone.

In the most seaward dunes and within the uppermost beach, the ghost crab (*Ocypode quadrata*), with its large, 1-inch diameter or so, circular burrow entrance, is a very common inhabitant. The burrow winds, curves, and recurves beneath the beach, sometimes reaching a total length of several feet. Sometimes the excavated sand arranged around the burrow entrance is a different color than the surface sand, indicating it originated in some buried sand layer.

The ghost shrimp (*Callianassa* species), which is common in South Carolina and Georgia beaches, builds its layer-disrupting burrows in the intertidal zone. These burrows have a diameter of around 1 inch, a knobby, clay-cemented exterior, and extend to depths of a few feet. This burrow wall is resistant enough to withstand low-energy swash, and sometimes the burrows stand in relief on the beach surface as chimney structures. Like the burrows of other organisms, sand fills the hole after the animal dies. Geologists can use the maximum elevation of filled-in burrows in Ice Age beach deposits, stranded on the lower Coastal Plain, to estimate the level of the sea in the past.

Plants such as sea oats (*Uniola paniculata*) and American beach grass (*Ammophila breviligulata*) extend roots deep into dunes, interrupting and

A crab hole with excavated sand of a different color than the beach surface, which indicates the crab's burrow cut through beds of different materials beneath the beach at Flagler Beach, Florida. Penny for scale.

Scour around a ghost shrimp burrow has left it standing as a chimney structure on the beach at Isle of Palms (South Carolina). The chimney is surrounded by an adhesion surface. Penny for scale.

distorting the characteristic fine-grained, thin-layered dune stratification. We discuss specific structures produced by plants and animals in the last section of this chapter.

## Bubbly Beaches: Holes, Blisters, Pits, and Rings

Beaches are naturally porous, and the pores between sand grains are either filled with water or air. In fact, a beach may have 35 percent or more open space, or pore space, between the grains, so a beach can hold a considerable amount of air and water. As the tide recedes, the water level within the upper few feet of the beach drops almost as fast as the tide. As water exits the beach, air replaces it, filling the spaces between grains. As the tide comes back in, water once again forces its way into the pore spaces, displacing the air and forcing it back out through the overlying sand.

On a smaller and faster scale, air and water exchange places as wave swash floods over and soaks down into the beach and then flows back to the sea. The mechanism of air release in this case is the weight of the swash water layer actually forcing air out of the beach. As the thin sheet of water that is the final gasp of a broken wave surges ashore, its weight forces air out of the upper few inches of the beach. The result is sometimes a virtual field of bubbles streaming through the water, especially on the upper beach in the

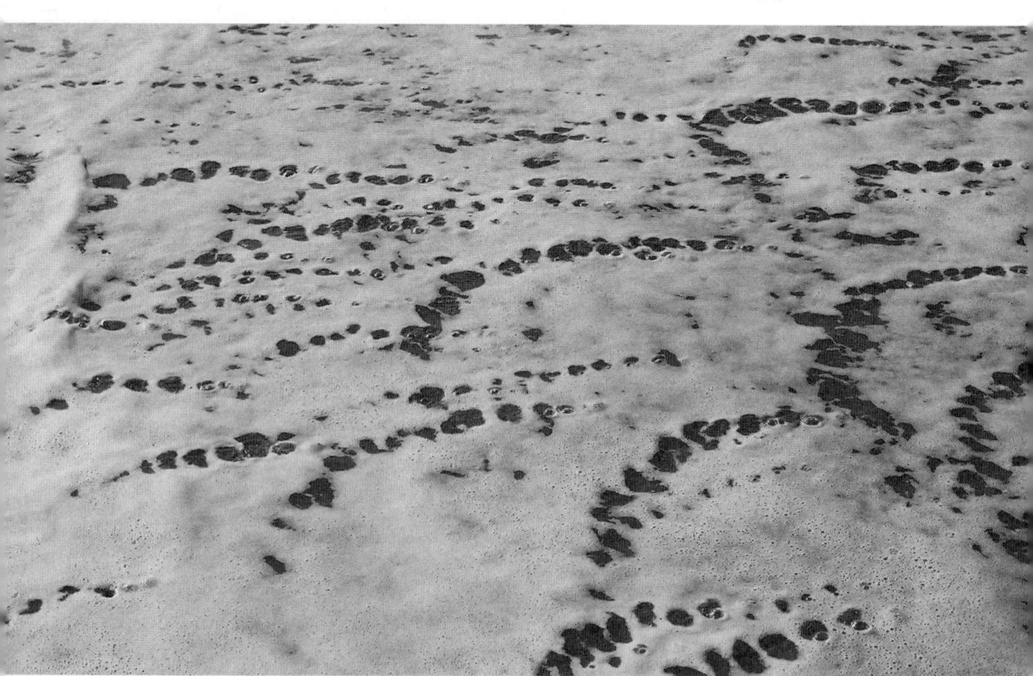

Air escaping the beach produces bubble trains in the foamy backwash at Sea Bright (New Jersey). The backwash was flowing to the left, so the air holes from which the bubble trains originated are located at the right-hand end of each train. The air holes may fill with sand or remain open as nail holes after the swash returns to sea.

rising tide. When the backwash retreats toward the next breaking wave, some air returns into the beach. Then the next wave swash arrives and the whole process repeats itself.

The late Sam Smith, an Australian engineer, hypothesized that another mechanism besides weight may be working within the thin layer of water that makes up swash. Smith based his opinion on the observations he made on a beach along the Gold Coast of Australia. He visited the beach at the same spot every single day, except when he was out of town, rain or shine, for over twelve years. On each visit he made and recorded a number of measurements of and observations about the beach and surf zone. Sam was a very observant individual, and as a result gained an extraordinary intuition about beaches. In some ways his little patch of beach in Australia may be the best-known beach in the world.

In a typical swash layer viewed from above, one can see small, clear patches of water that are a foot or two in diameter and surrounded by white lines of ephemeral foam and bubbles. The best way to look for these swash "footballs," as Smith called them, is to observe the swash zone from a fishing pier. You can see them from the beach, as well. The swash footballs aren't always there; if they are, they will be especially visible in the swash that covers the flatter portions of a beach.

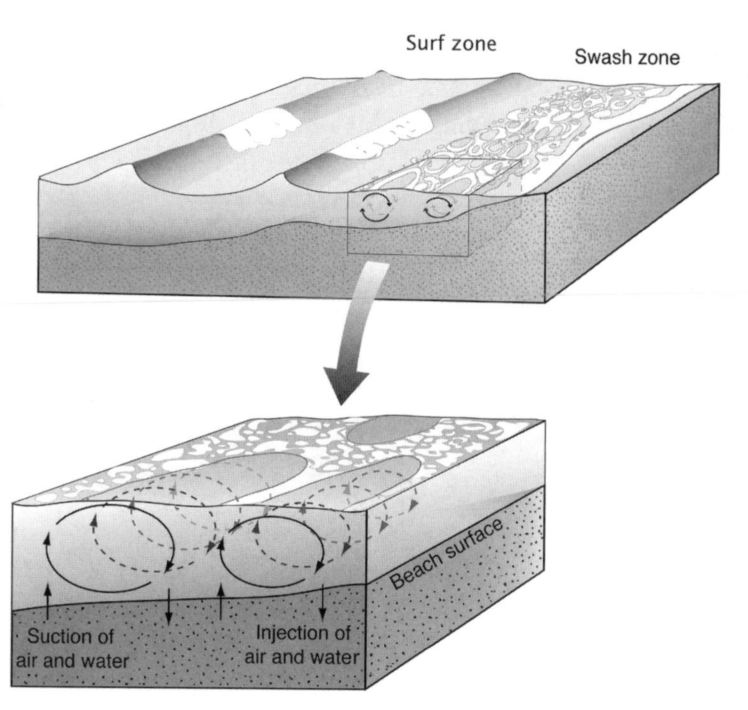


Diagram of how swash "footballs" may force air in and out of the beach. —Drawing by Charles Pilkey

Each of these footballs is undergoing orbital motion, which it inherited from the larger orbital, or circular, motion of a breaking wave. Smith believes that as this orbital crosses the beach, one part of the orbital is ever-so-gently forcing water into the beach while the back side of the orbital is pulling water and air out of the beach.

As noted, to understand surf-to-swash processes on a beach, we recommend viewing the beach from a pier. You can watch the transition from swell to breaking waves, observe types of breakers, and see the foam-outlined "football" orbitals developing in the swash zone. We've seen particularly well-developed swash footballs from the municipal pier at Folly Beach (South Carolina). This is not just a fishing pier but also a strolling pier, and it is a particularly well-suited and pleasant platform from which to learn about the surf zone. Most fishing piers up and down the East Coast (most are in the Carolinas) will let nonfishing individuals on the pier for a small price to observe waves. Unfortunately, fishing piers are a vanishing species as they are removed by hurricanes and replaced by condominiums.

Whatever the mechanism, falling and rising tide, swash moving back and forth, or swash footballs within the surf swash, a lot of air is forced in and out of the beach, forming a wide variety of features. The air comes out of point sources through a limited number of holes, rather than being released through the sand uniformly over a broad area. You can easily see air coming out of a beach if you stand near the uppermost reach of the swash zone. Look for streams of air bubbles passing through the thin layer of swash. The bubbles form a line, or bubble train, as the swash carries the newly formed bubbles away from their hole of origin, either up or back down the beach.

The amount of air released varies from beach to beach, and on the same beach from time to time. There are even big differences in the rate of air release at different places on the same beach. It makes sense that the greatest amount of air released from a beach would occur at high tide or on the upper beach, when and where the full range of the tide pushes air up through sand grains. Bubbles are common in the swash layer, and the surface features they form are most commonly seen on the exposed beach just above the swash zone.

Sometimes the upper beach surfaces are virtually full of holes, giving them an almost sievelike appearance. These holes range in diameter from that of a very small tack to about a sixpenny nail, and their depth ranges from a tiny fraction of an inch to over an inch. Since their diameters are comparable to hardware nails, we call these features *nail holes*, although their formal name is *sand holes*.

Nail holes occur on lakeshores, which is evidence that the swash zone itself can produce holes without a tide. A well-developed line of nail holes and associated features can often be seen on lake shorelines where wind or storm waves, not tides, change the water level.

After storms, or on other occasions when "new" bodies of sand are deposited on the beach, air-related features seem to flourish. New sand is capable of storing a lot of air and is particularly susceptible to the formation of nail holes and other air-release features. As the sand resides on the beach for awhile, a few weeks or months, we think that the grains settle in and are packed

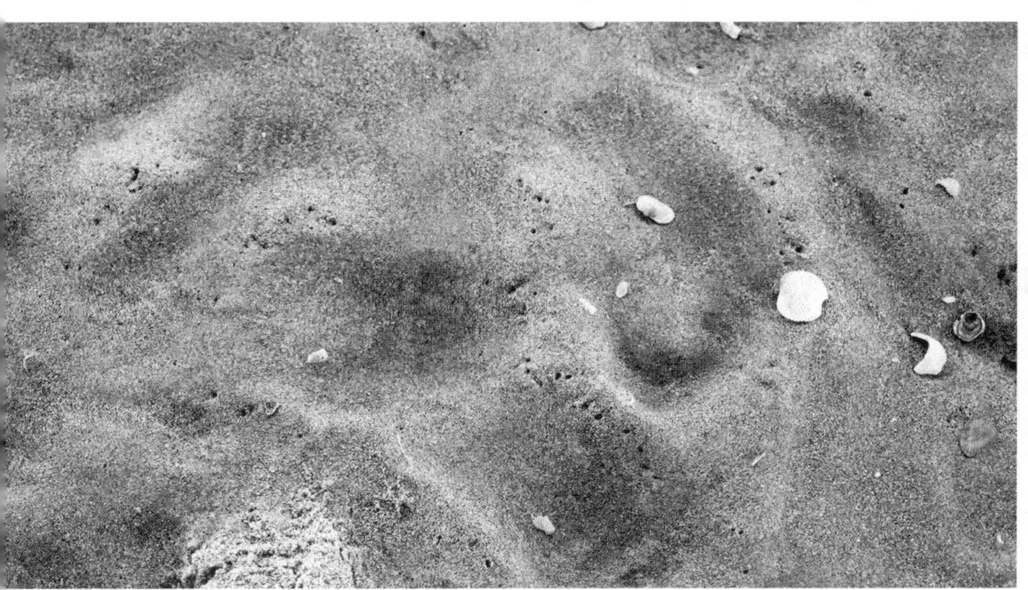

Nail holes concentrated around the edges of footprints that swash washed over. The disturbed sand around the edges of the footprints held more air between the sand grains than the compressed bottoms of the prints.

Nail holes vary in size and distribution. Most of these holes were formed by escaping air. Similar holes can be produced by burrowing organisms, such as the two larger holes to the right of the penny.

more tightly as the sand compacts and air moves less efficiently between sand grains. Therefore, the older, less porous sand stores and releases less air than fresh, newly arrived sand. Tire tracks and footprints can rearrange the packing of sand, and swash breaking over tracks and prints can cause a lot of holes to form in the loose sand around these disturbances. Holes are particularly concentrated around the margins of footprints.

There has been little scientific study of air-formed beach features, although they are occasionally mentioned in the geologic literature. Our general discussion of these features is based on years of walking beaches, but we have never systematically investigated them. Our discussion of air in the beach only scratches the surface, so to speak, and much remains to be learned.

We recommend that you put down your beer, binoculars, fishing pole, or the novel you brought to the beach and make some close-up beach observations. You could count the number of sand holes or other features, such as volcanoes or rings per square foot, and note their diameters. Do this at different stages of the tide, before and after storms, on different parts of the beach, or make repeated observations at the same spot. You could watch these structures form as you stand ankle deep in the swash zone. However you do it, you will have the satisfaction of learning something about the beach that no one else knows!

The simplest, and by far the most common, surface expressions of escaping air are nail holes. If you take the time to count them, nail holes occur in concentrations of 5 to over 150 per square foot. The densest concentration of holes we have seen, greater than 200 per square foot, was in South Beach section of Miami Beach, Florida, an artificial beach. In coarser sands, such as those in some New England beaches, and in gravelly sands, both shell gravel and pebble gravel, nail holes are generally sparse or absent. This is likely because air can easily pass through the coarse materials of these beaches without creating passage holes.

When the volume of air is larger than that required to form nail holes, sand volcanoes may form. In this case, both air and water escape rapidly enough to carry sand out of a hole. The sand accumulates around the hole, like the cone of a volcano, albeit a miniature one. Sand volcanoes don't stand up well if swash washes over them, so these features are preserved at the highest point of the swash where they won't be disturbed, at least until the next tidal cycle.

Sometimes the surface layer of the beach, a few sand grains thick, is cohesive enough to trap a pocket of escaping air, which causes this layer to bulge upward rather than form an escape hole to the surface. These circular

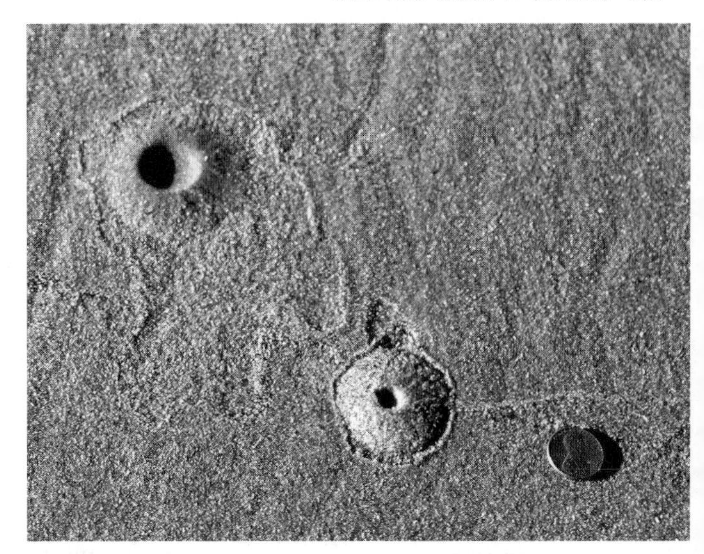

Sand volcanoes on the beach at Isle of Palms (South Carolina). Penny for scale.

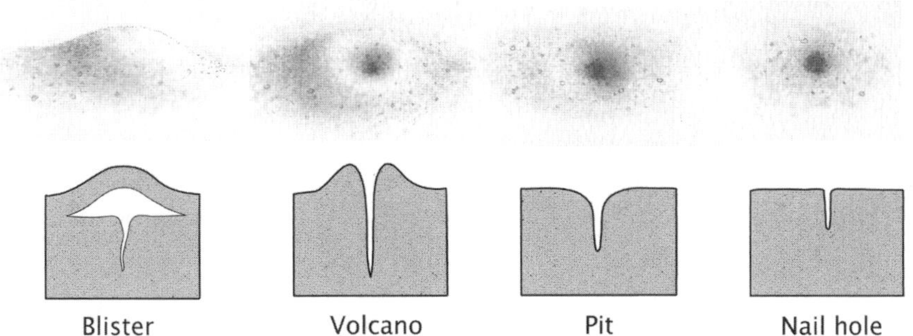


Surface views and cross sections of various types of cavities and holes produced by the entrapment and escape of air in beaches. Blisters form where air accumulates to form a cavity that forces the surface of the beach to rise slightly. Sand volcanoes form when escaping air and water carry sand grains out of a hole and the particles accumulate and resemble a volcanic cone. Pits form where swash decapitates the top of a blister or sand volcano. Nail holes are small sand holes of variable size that form where air escapes. —Drawing by Charles Pilkey

mounds or bulges are termed *blisters*, and they are somewhat analogous to a cheese soufflé in cross section. Most of these bumps extend upward from the plane of the beach only a fraction of an inch, but some are larger. Most are at least the size of a quarter, and some are fist sized.

The best blisters we've seen have been on beaches with fine sand; for example, those at Coney Island, New York, and Daytona Beach, Florida. This suggests that finer sand may form the crust of a blister due to its greater cohesiveness, but there are other explanations as to why the upper layer of a beach may be cohesive. An almost invisible organic film made up of algae or diatoms may cause it, or a very light layer of salcrete.

Sometimes blisters occur in fields, where the individual blisters are surprisingly symmetrical, or evenly spaced. These fields, which can be tens of feet long and several feet wide, usually form in a very uniform, fine sand. The symmetrical arrangement of the blisters is a puzzling sight to beach strollers alert enough to notice them. Some people suspect the hand of humans in such symmetry, but this is not the case. We suspect these blisters develop as air moves uniformly through a layer of uniform sand in the upper beach, where the even spacing of the blisters represents an efficient pattern of air release through the beach sand.

Sometimes when a blister is planed off, or decapitated, by swash, a pit is left behind. The size of pit depressions is determined by the size of the eroded blister, but they are generally larger than sand volcanoes. With a little effort on hands and knees, you can dissect blisters with a pocketknife or a kid's sand shovel. A cut through the center of the blister reveals a dome-shaped pocket of air that may be connected to a nail hole, from whence the air escaped.

Under just the right circumstances, blisters form one of the more spectacular bed forms on the surface of a beach. If the thin layer of semiconsolidated sand that bulges up to form the blister contains a heavy-mineral layer, a ring will form on the beach when swash decapitates the blister. Often the air pocket that formed the blister is filled with sand as the blister is being planed off, leaving a perfectly circular ring, ½ to 3 inches in diameter. A field of

Pits along this swash line on Assateague Island (Maryland and Virginia) formed as blisters were truncated. Some of the blisters still persist left of the penny and between the two swash marks. Some larger nail holes are evident as well.

Cross section of a blister in fine-grained sand. Note the nail hole in the top of the blister. Small pits where other blisters were truncated are visible. The coin is about the size of a nickel.

The blister to the left of the penny and a few large nail holes at Sea Bright (New Jersey) produced the bubble trains pictured on page 132. Note the strong lineation from the upper right to lower left that backwash produced. Penny for scale.

This broken crust of the beach at Cedar Island (Virginia) is an example of salcrete. Salcrete forms when salt from seawater crystallizes between sand grains, forming a weak cement that holds the sand together as a crust. The dark color is due to an abundance of garnet sand grains. Penny for scale.

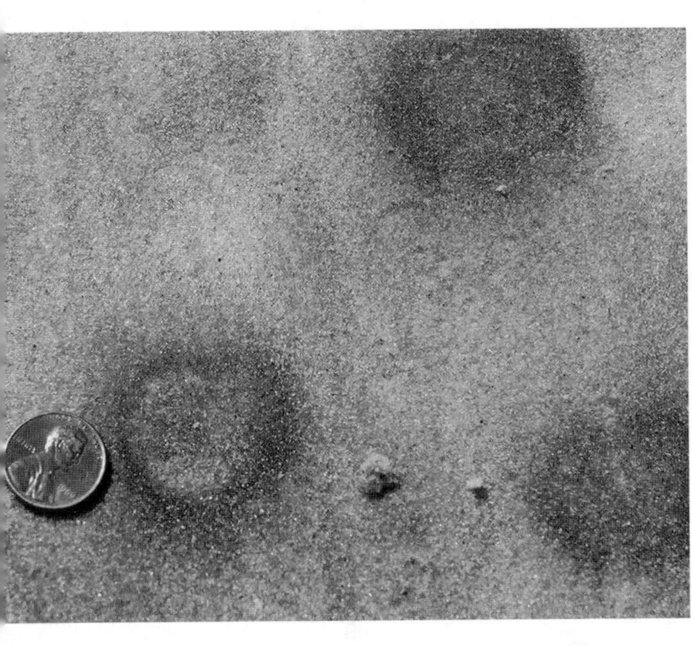

Ring structures form when swash truncates a blister. If heavy mineral layers are present, the rings will be more apparent. Penny for scale.

blisters can turn into a field of evenly spaced rings. Whether a field of evenly spaced rings, or a lone ring structure, these features are unforgettable.

Rings can develop on any beach with fine to medium-sized sand, but they are probably best developed in the beaches of Georgia, the Carolinas, and Virginia that are relatively rich in heavy-mineral sand. Most South Florida beaches lack enough heavy minerals to develop pronounced ring structures, so rings are less common there. Rings also are found on nourished beaches. Look for rings on Cape Cod (Massachusetts), the Outer Banks of North Carolina, Hunting Island and Folly Beach (South Carolina), and Jekyll Island (Georgia).

### *Make Your Own Structures*

You can try to manufacture your own nail holes, volcanoes, or with luck, a blister or two. Make a deep footprint and then observe as swash falls over the edge of the depression. If conditions are right, the water cascading over the lip can result in a line of air holes, volcanoes, or blisters at the rim of the print. The pressure of the infiltrating water forces air out of the looser, disturbed sand around your print.

Sometimes placing a piece of driftwood at the uppermost limit of the swash zone will create conditions that allow nail holes to form. As water rushes over the obstacle and onto the beach on the other side, a line of bubbles may form next to the stick. There is yet another way. Stand still for a moment on the wet beach and watch the sand within a few inches of your foot. Shift your weight down on one foot. You may see nail holes appear

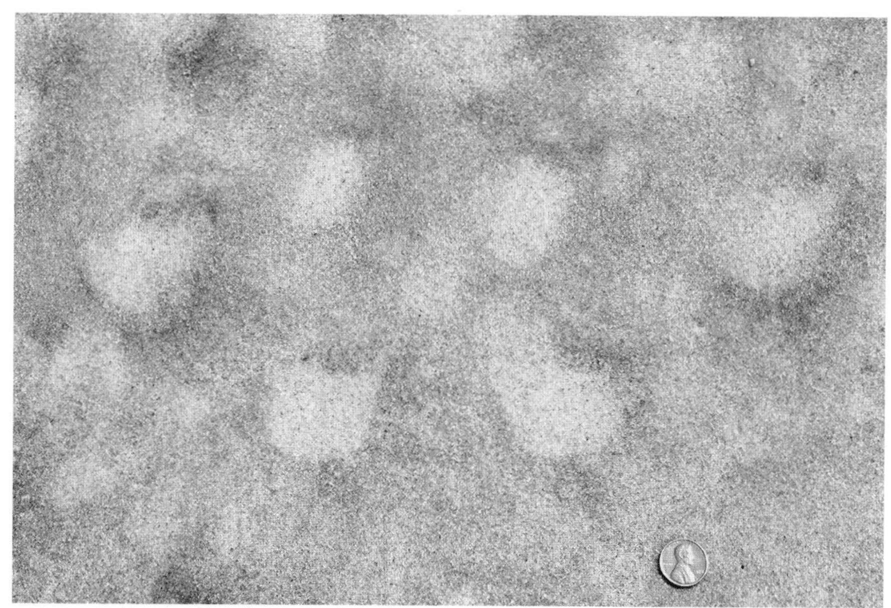

Ring structures often occur in patches, as seen here on Pea Island (North Carolina). The rings may be evenly spaced and of uniform size, or random in size and distribution. Penny for scale.

around your foot because the pressure squeezed air out of the beach. We first observed this phenomenon on the beach at Atlantic City, New Jersey, which is a beach with fine sand, but the result should be the same for any sandy beach with a high percentage of pore space. If it doesn't work the first time, try it at other locations on the beach.

### *Holes and Critters*

Most beach strollers assume that sand holes are produced by critters of some kind. But most holes, blisters, pits and rings are physical features caused by escaping air. The same is true for some sand volcanoes. The observant beach stroller can even see these features form as noted above.

Critters do use these holes. On hands and knees, we've observed sand fleas jumping in and out of nail holes on the beach at Sandy Hook (New Jersey). The fleas were occupying previously formed holes. But critters do form holes that may be somewhat similar to air holes, although they are usually deeper. We discuss burrow holes in the beach produced by animals later in this chapter.

The bottom line is that most sand holes in beaches, along with blisters, and some sand volcanoes, are formed by the action of falling and rising tides and the pumping action of the swash zone. Just how important critters may be in producing similar features can best be answered by spending some time on your hands and knees observing, excavating, and dissecting the holes and related features to determine their origin.

## Knee-Deep in Sand: Soft Sand

Have you ever been walking along on a solid beach when suddenly the going got tough? Your feet sank into the sand and every step took extra effort? You were in soft sand. If this molasses-like consistency continues for a few hundred yards, a pleasurable beach walk can become the hike from hell! We've experienced this on Georgia beaches, where one may sink knee-deep into beach sand!

The soft sand that mires beach walkers is called *bubbly sand* or *cavernous sand*. Stretches of soft sand occur where escaping air has formed cavities in the beach, giving it a very porous fabric. As you walk on the sand, the underlying holes collapse and your feet sink into the sand.

Bubbly sand is formed by the same beach-breathing phenomenon that is responsible for nail holes. Bubbly sand is found in layers and usually extends only a few inches into the beach, but it can extend down as far as 1½ feet. The bubble cavities in a layer may make up one-third of the layer's volume, and these cavernous spaces within the sand range between ⅛ and ½ inch in diameter. Occasionally, the bubbles have diameters as large as a quarter. In the uppermost layer of a beach, the bubble holes are close to spherical in

Footprints in soft sand at Sandy Hook (New Jersey).

shape, but with depth the spheres are flattened somewhat by the weight of overlying sand. Forcing your hand into the beach and bringing out a chunk of sand will reveal the bubbly texture of the sand in the fresh fracture face of your sample. Bubble holes can also be seen on the face of a small ditch carefully dug across the beach, or where a natural channel has cut through the beach.

Bubbly sand forms mainly in well-sorted, fine to very fine sand that contains little shell material. It also tends to form on beaches with high tidal amplitude. Georgia beaches have, on average, the finest-sized sand and the smallest amount of shell material on the East Coast. They also experience relatively large vertical distance between tide stages (7 to 11 feet), so bubbly sand is well developed.

A wedge of soft beach sand shows its open, porous structure. The entrapment of air within the beach produces this cavernous, or vesicular, porosity and results in either a lack of laminae or poor development of them.

Bubbly sand has a fabric that resembles Swiss cheese in cross section; trapped air forms the holes in the sand.

In a very general way, tidal amplitude increases in a southerly direction between the mouth of Cape Henry (Virginia) in Chesapeake Bay and the Georgia Coast, and grain size of beach sand decreases in the same direction. As a consequence, one may experience shoe-deep soft sand in North Carolina, more than ankle-deep sand in South Carolina, and occasionally knee-deep bubbly sand in Georgia. But as with other generalizations about the location of beach features, bubbly sand is where you find it. Imagine driving a vehicle through knee-deep soft sand. More than one hapless Georgia pickup-truck owner has watched the tide flood his vehicle as it was firmly stuck in bubbly sand. And bubbly sand can't be fun for nesting sea turtles either.

Bubbly sand is found mainly on the upper beach, usually above the midtide mark. Most likely, bubbles form as the rising tide forces air to rise up through the beach. Because bubbly sand forms in very fine sand, air does not flow as readily through it to the beach surface to form escape structures, such as nail holes, as it would on a somewhat coarser beach. Wave swash from incoming waves also inhibits air escape. The air can't move down and it can't move up, so the pressure forces bubbles to form in the layer in which the air is trapped. The air cavities probably grow larger in succeeding tidal cycles as they repeatedly drown and dry out.

A vehicle that encountered bubbly sand on Currituck Bank of Virginia and North Carolina. Otherwise firm beaches can have patches of soft sand that offer such surprises. —Photo by Drew Wilson/*The Virginian-Pilot*

Under rare conditions, soft sand can be dangerous to beach walkers, especially children and those who might have difficulty walking. Like all beach features, soft sand comes and goes, so you can never be sure where you will encounter it. We know of an instance at the north end of Kiawah Island (South Carolina) where a person riding a bike on the beach suddenly rode into soft sand in an area that also had mud layers. The rider had a difficult time extricating himself, crawling through the sand and mud and abandoning his bicycle to the ravages of seawater. Bubbly sand of this kind is a rare occurrence, but beachgoers should be careful.

## Leaky Beaches: Groundwater in the Beach

Barrier islands are completely surrounded by saltwater. Yet the first European settlers on these islands dug wells and hit shallow freshwater in relatively large quantities. This may seem contradictory, but fresh groundwater is common under land immediately behind beaches, and this subsurface water supports the plant and animal life on barrier islands. Groundwater also sustains the human population. It may even play an important role in the erodibility or stability of the beach. The presence of groundwater in the beach is usually only obvious near the low tide stage when freshwater begins to escape from the lower beach.

Water gets into the land behind the beach and under barrier islands through the infiltration of rain and snowmelt. Pour a bucket of water on the dry beach or on a dune and watch infiltration happen. The water disappears in a flash, but where does it go?

The water passes between the grains of sand and continues moving downward until it is stopped by some sort of barrier. Under a New England beach, that barrier often is bedrock. Under barrier island beaches, layers of mud, peat, or the top of the zone of sand that is saturated with saltwater block the water's downward flow. Freshwater is less dense than saltwater, so it is not able to penetrate or mix with the denser, salty fluid. Just as oil and water mix poorly, saltwater and freshwater resist mixing. Whatever the obstacle, when freshwater can no longer move down through the pores, it accumulates and fills in available pore space.

The surface of this accumulated underground water, whether fresh or salty, is called the *water table*. The upper surface of the groundwater under the uplands near beaches is above the level of the high tide line. Therefore, groundwater flows laterally and downhill toward the sea. So fresh groundwater takes on the shape of a lens, thicker in the center of an island or dune field and tapering out and downward to the edges of the shoreline.

The elevation of the water table and thickness of the freshwater lens depend on the amount of local rainfall, local rates of human consumption, and outflow. During really rainy periods, some of the lower areas between sand dunes may flood to form small ponds. The surface of the ponds represents the top of the local water table, which has risen in response to heavy precipitation. Usually, however, the surface of the freshwater lens is deeper, and dune plants, especially grasses, have relatively long root systems to tap groundwater even during dry summers.

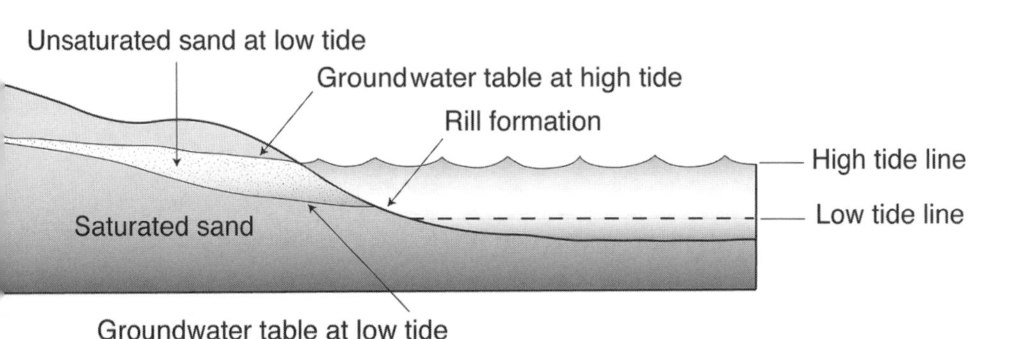


Cross section of a beach showing the generalized location of the water table. Where the water table encounters the ground surface, water flows out as a spring. Seepage from the beach will be most apparent at low tide. —Drawing by Charles Pilkey

A spring emerging on a Maine beach at low tide.

People have become major consumers and producers of beach groundwater. Plants may naturally conserve water, but people often take abundant water for granted. Wells dug to draw water exclusively from the shallow groundwater soon go dry under even moderate rates of consumption. As more people come to beaches, water must come from new sources, either deeper wells into bedrock or Coastal Plain sediments or pipelines from the mainland.

Since groundwater flows downhill and out to the beach, the lower the tide the greater the slope of the water table, and the more likely water is to flow out of the beach. The intersection of the water table with the surface of the earth is called a *spring* or *seep*. Springs of many sizes and shapes occur on beaches. Similarly, beachfront houses with leaking septic systems may create their own lenses of groundwater that seep out onto the beach.

Rill marks, which are small erosional channels, are the most visible sign of groundwater escape. As the tide falls, groundwater oozes from the sand and flows down the slope of the foreshore. As it moves it carries sand grains

Where large amounts of groundwater discharge from the beach at low tide, a stream may form. In this example, the velocity of flow was great enough to form a shallow channel with current ripple marks.

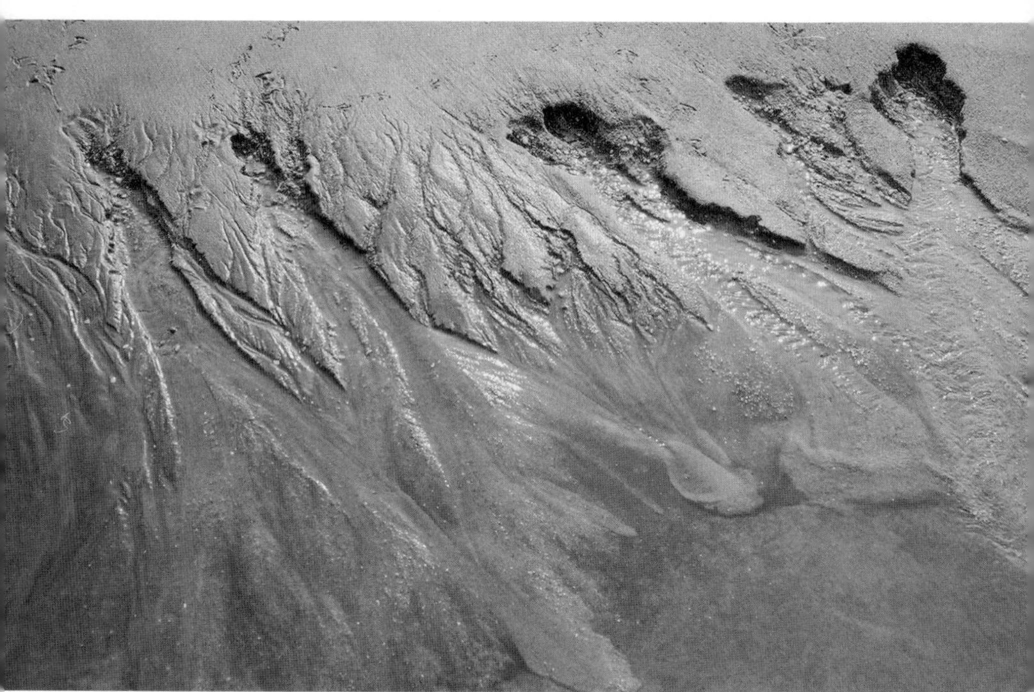

Rill marks forming on North Beach at Sandy Hook (New Jersey) with their associated rivulets and microdeltas. Groundwater flow from the beach creates a miniature river delta system. Scarps at the head of rills erode and migrate in a landward direction.

with it and forms small channels. Tributary "streams" from other seeps flow into a main channel, and as low tide approaches, something resembling a tiny river system becomes visible. The sediment carried down the channel of a rill is deposited as a tiny microdelta at the end of the rill. Amphitheater-like scarps at the landward (headward) end of rills form where the groundwater escapes. Where rills are closely spaced, these terminal fans will join, somewhat analogous to the coalescing alluvial fans at the foot of a mountain range in a desert. If you find a set of rills forming, watch them for a while as their tiny channels and fans change in size and shape—a sped-up micromodel for how larger natural streams evolve.

The size of rill marks is highly variable, ranging from a few inches to several feet in length and width. Some beaches have none, while others have many. Rills are more likely to develop during periods of heavy rainfall and maximum spring low tide than in dry times or at maximum high tide. The shape of the beach may also play a role. Steeper beach surfaces produce more rills because the seeps will flow faster and cut through the slope face. Rills don't form on flat sections of beaches, but at low tide you may notice that wide flat beaches have a mirrorlike surface from being completely saturated with groundwater.

If you dip a finger into the water running down a rill and taste it, chances are good that the water will be salty. The water seeping out of the beach is usually seawater that the freshwater behind it is forcing out. Just before the rising tide once again covers the rills, the water may become fresh.

The numerous salt ponds of New England provide evidence that water moves through the beach even though we don't always see it. These coastal ponds contain fresh to brackish water that varies in saltiness during the year; typically they are not connected to the ocean by a permanent inlet. A small inlet may occasionally be opened, naturally or artificially, to allow water exchange between the pond and the sea, but most of the exchange is groundwater related. When no inlet exists, the water level of the pond is usually higher than the ocean during low tide. Pond water then seeps through the beach. Along the south coast of Long Island, in the vicinity of the Hamptons, there is a series of coastal ponds (Mecox Bay, Sagaponack Pond, Wainscott Pond, Hook Pond) that are blocked by small baymouth barriers. Lots of nutrients from septic systems and surface runoff results in algal blooms. Water quality is actively managed by seasonally opening inlets with a backhoe. Similar settings and problems exist along the south shore of Cape Cod.

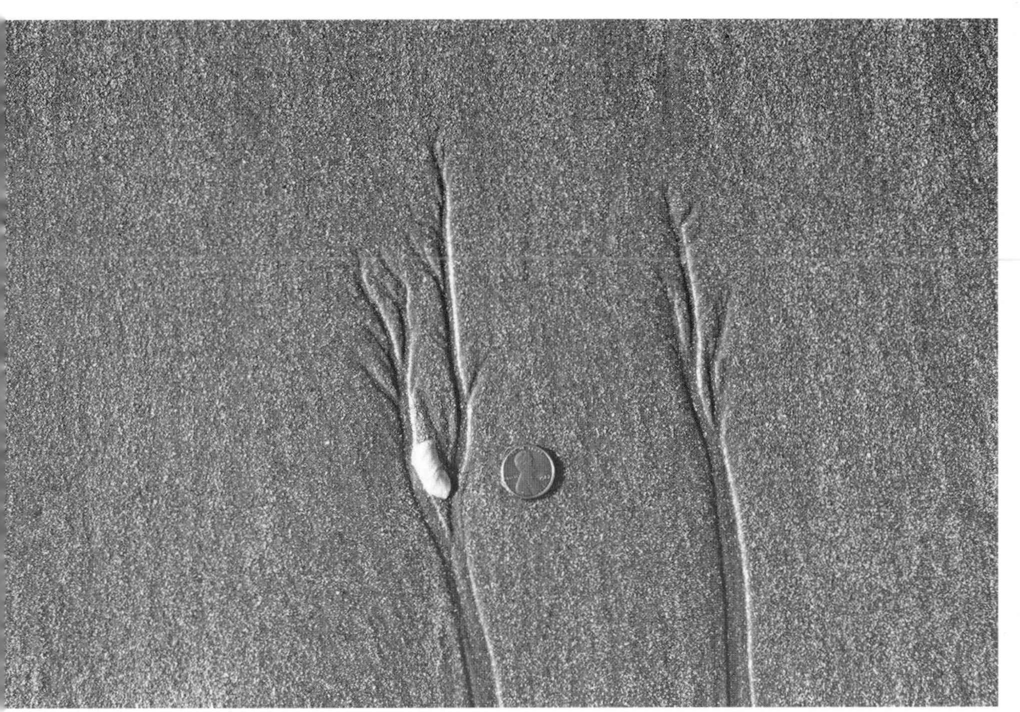

A close-up of the head of a rill shows its dendritic pattern of erosion. Penny for scale.

Water in the beach affects how sand is moved about. If surface sand is saturated, not much water from a wave sinks into the beach. Instead, water from the wave rushes back to the sea, carrying beach sand back to the sea as well. This process tends to flatten beaches. The constant movement of groundwater through the sand to the sea creates a small force that pushes the grains apart. This pore pressure weakens the sediment, and geologists assume that this makes it just a bit easier for waves to pick up sand and move it about.

To reduce the sand erosion caused by waves, engineers have attempted to take advantage of this principle and tried to plumb eroding beaches, for example, at Nantucket Island (Massachusetts). A perforated pipe is buried in the beach and groundwater is pumped out. In theory, this creates a reverse flow of seawater into, rather than out of, the beach. This inflow is supposed to make it more difficult for waves to remove sand grains, and it may even enhance sand deposition. Ideally, this might work, but storms quickly overload and erode the plumbing. Maintenance of the pumps is costly because of saltwater corrosion. And then there is the sound of engines on the beach. All of us who like the serenity of beaches hope that plumbing does not become a popular beach-preserving fad anytime soon.

## Squeaky Beaches: Barking and Singing Sand

Each of the authors recalls the first time he encountered singing beach sand, that high-pitched squeak or bark that is produced while walking on some dry beaches. Perhaps you've had the same experience while walking on a beach but didn't give the yelps or chirps a second thought. Typically, small children are the first in a group to notice these squeaking sounds, either because their hearing is more acute or they're better tuned-in to nature's sounds. For playful parents, singing and barking sand can be the source of a number of games, from imagining creatures making the sounds to imagining that sand grains are complaining about being stepped on.

The phenomenon, long recognized in the technical literature, goes by a variety of names, including barking sand, musical sand, singing sand, sonorous sand, sounding sand, squeaking sand, whistling sand, and in Japan, frog-sound sand! The yelp is produced by the shear that occurs when sand grains slide against each other as you scuff your feet on the surface.

Singing and barking sands occur on most sandy beaches of the world, and U.S. Atlantic beaches are no exception. Henry David Thoreau was one of several nineteenth-century naturalists who recognized the phenomenon on American beaches, noting singing sand in the Manchester, Massachusetts, area as well as at Cape Cod. Manchester's beach came to be called Singing Beach, and similar squeaky beaches are found in Massachusetts on Plum Island and along the coast of Cape Ann and Gloucester. During Thoreau's time, musical sands were of considerable interest to pioneer geologists. One of them, H. C. Bolton, published a list of more than sixty locations of singing beaches from Maine to South Carolina. Some of his studies focused on Rockaway and Far Rockaway beaches on Long Island (New York).

Sand deposited by either water or wind may bark or sing. The pitch and volume of the sound is in the ear of the beholder, but they do differ considerably from beach to beach and from time to time on the same beach. For sand to bark or sing, a number of conditions must be met. Typically, these sands are fine to medium in grain size and well sorted—that is, all of the sand grains are of similar size. Fine material, such as silt or clay, should be absent, and a high percentage of grains should be spherical in shape with smooth and dust-free surfaces. Sands with a high quartz content typically have such properties. A beach composed of mostly very fine carbonate sand (shell fragments) with little quartz, such as Miami Beach, isn't as likely to sing. Shelly sand is not supposed to bark, and medium-sized sand is supposed to bark louder than finer sands. We use the qualifier "supposed to" because very little is known about such sound-producing sands even though geologists have long been aware of the phenomenon. For example, we have encountered shelly sand on Folly Beach (South Carolina) that was poorly sorted, and yet it barked. So much for one widely held perception about this phenomenon.

Fundamentally, the most important determinant of singing or barking is the way the sand grains are arranged, or packed. Singing sands often have a rhombic packing pattern, which means each layer of grains is offset from the one below. When you step on the sand, the grains are forced to move down and out past each other and into a new packing arrangement. This action generates the shear that produces the sound. Geologists think that friction between grains is reduced and the sand is silent when dust or other small particles are on the sand grains, or when the sand is poorly sorted. Because beach and dune sands are washed or well sorted by waves and wind, they typically are dust free.

The squeakiness of beach sand may vary seasonally. The best sound occurs in spring after winter storms have re-sorted the sand and removed the finer sediment. A Japanese study of beaches suggested that the loss of the ability to sing may be an indication of pollution. The accumulation of very fine organic matter between sand grains has the same effect as dust, reducing the sand's ability to generate sound.

Not all sands are of equal voice, and a large frequency range occurs between different sands and localities. Fine sands may produce only a poor, weak squeak, while medium-sized sands can emit a range of sound, from faint to sharp. The clearest sound usually comes from dry, medium-sized sands at the back of the beach. A vehicle driving through such sand can produce a continuous grinding sound that is particularly loud.

Water also is a factor in the ability of sand grains to produce sound. Although water usually silences singing sands by causing the grains to stick together instead of shearing past each other, adding small amounts of water to the sand, such as a light rainfall, can actually raise the pitch of the sound. So sand primarily sings on the dry upper beach above the normal high tide line, but some sands have been reported to sing on the lower beach near the low tide line. There is still a lot to learn about barking sands.

### *Booming Sand*

While beach sands usually sing with a very brief, high pitch (500 to 2,500 hertz), the bass of the chorus is booming sand, which refers to the sound produced as sand slides down the face of large dunes. The booming sound is louder and of lower frequency (50 to 300 hertz) and has been described as a low-frequency "roar" or "groan." It can be quite startling, especially at night! Booming is primarily a desert phenomenon and is even called *song of the desert*; however, coastal dunes occasionally produce the noise under dry conditions. The booming may persist for several seconds, or, more rarely, for several minutes in places like the great sand seas of the world (for example, the Sahara and Gobi Deserts). The sound may even fluctuate in volume.

Descriptions of booming sand occur in historical records as far back as ninth-century China, where people slid down the Mt. Ming-sha-shan sand dunes to announce the arrival of a boy's festival day with thundering sounds. Marco Polo attributed the desert sounds to musical instruments, drums, and the clash of arms produced by evil spirits, not unlike the folk tradition on Kauai in the Hawaiian Islands that regards the booming from dunes as the sound of disgruntled spirits. We know of no reports of booming sands on East Coast dunes. We suspect that booming could occur on large dunes, such as those around Nags Head and Kill Devil Hills, North Carolina, or Cape Cod (Massachusetts), particularly during extended dry periods.

## Leave Only Your Footprints: Burrows, Tracks, and Trails

"Leave only your footprints" is a popular environmental slogan and very appropriate advice for people recreating and living along Atlantic Coast beaches and dunes. Unfortunately, as noted in the earlier discussion of wrack, often the dominant traces of organisms on beaches are the varieties of trash that we humans leave behind. The plants and animals that inhabit a beach or dune leave traces that are as fascinating as the physical structures of those beaches and dunes.

We see the traces—tracks, trails, burrows, and fecal material—of animals more frequently than we see the animals that produce them. For example, the trace of a crawling turtle is a feature you might see in summer on a southeastern U.S. Atlantic beach, but the turtles come ashore at night when few people are about to see them. Their distinct trail in the sand, however, tells the story of their nocturnal nesting visit.

A turtle returning to the sea leaves its familiar tank-tread trail across the beach. —Photo courtesy of David Godfrey, Caribbean Conservation Corporation & Sea Turtle Survival League

A close-up of a pit in a beach shows burrow structures; the burrows, filled in with lighter sand, crosscut the horizontal layers.

Some traces, such as burrows, are obvious, both on the surface and in cross section, while others may pose a mystery. Recreational beaches at the height of the season may be so worked over with human footprints that the surface resembles a ploughed field, but even on these beaches not all traces are obliterated. Find a clear area of beach or dune surface and you may discover a variety of life traces. Or dig a hole and create a vertical face in the beach, in which you will find ample evidence of its fauna.

### *Beach Critters: Little Ones and Big Ones*

Beaches are a tough place to live. Creatures have to cope with high-energy waves, wind, currents, and alternating wetting and drying. Nevertheless, beaches are alive. When you spread your towel on the beach, most likely you are in the midst of millions of animals, but you can see only a few of the largest. Biologists divide beach animals into two categories: meiofauna, the microscopic organisms that we probably don't realize are in the sand, and macrofauna, the animals big enough to see even though some are pretty small.

The tiny animals of the meiofauna are so small that they live in the pore spaces between sand grains of the beach, perhaps on the order of a million individuals under a square yard, often of the same species. How small does a critter have to be to qualify as meiofauna? Individuals can move through the

A mystery structure in a cross section of a beach. What caused some of the laminae to be folded, rather than horizontal? See the end of this section for answer.

sand without displacing sand grains. That means they do not produce burrows or other traces of their presence. Since they leave no visible footprints, tracks, or trails, you need a microscope to see evidence of these abundant animals. Regardless, when you scoop up a handful of sand and look at the sand grains, you are also holding a collection of critters. Some meiofauna is visible to the naked eye, but you may not know what to look for. An abundance of microscopic algae and diatoms gives a wet beach a slight yellowish or greenish tinge, which is most noticeable on a cloudy day. Some organisms of the meiofauna are phosphorescent, and if you take a night stroll on the beach, stomp on the sand, or drag a stick through the sand you may see them glow.

Although diverse species are present on the beach, nematodes (microscopic worms) typically make up to 85 percent of a beach's meiofauna. These animals are important to the beach because they feed on bacteria, algae, organic detritus, and dissolved organic matter. In short, they clean the beach as they move incessantly between the sand grains, in both the air- and water-filled pore spaces. As every beach bird knows, the meiofauna is also important to the food chain. Just as these microorganisms are food for beach birds, they are also an important food source for animals that inhabit the sandy nearshore zone. One common meiofaunal form, *Harpacticoida* species, has been found in the stomach contents of bottom-feeding fish. So an abundant and diverse meiofauna may be an indicator of the overall health of a beach. Placing dredged sand on beaches and beach bulldozing, of course, destroys the meiofauna.

### Beach Traces

The macrofauna includes beach creatures that leave tracks, trails, burrows, and fecal material that tell us of their activity. The high energy of the beach environment favors burrowers that can take shelter below the surface. Burrowing is also a feeding adaptation. Sometimes you will find a trace an animal produced before it burrowed beneath the surface, and what created the trace may remain a mystery.

While vertebrate animals, such as birds, are common and easy to identify from their footprints, invertebrate animals occur in greater diversity in the beach environment, although only a few species are usually dominant. A study on Sapelo Island (Georgia) revealed fifty invertebrate species in the tidal flat environment, but less than a dozen species accounted for 85 percent of the animals present; they were mainly polychaete worms, crustaceans, and mollusks.

The presence of a polychaete worm may be apparent at the surface if its burrow is lined with unique material and the upper part of it is exposed. For example, the *Onuphis* worm lines its burrow with a parchmentlike material that it secretes, and sometimes these small tubes are exposed on the lower beach at low tide. The genus *Balanoglossus* constructs a burrow that is U shaped in cross section with a funnel-shaped opening on one end and a ribbon of extruded sand that looks like toothpaste squeezed from a tube. Other worm tubes, such as those of the genus *Diopatra*, are constructed of a hard outer material that incorporates shell fragments and plant detritus; they may stand in relief on the beach surface as waves scour sand from around the top of the burrow tube.

If you want to see some small animals, like polychaete worms, get a piece of window screen or a kitchen strainer and wash beach sand through your sieve. Samples from the lower beach may yield some animal specimens that you normally don't see, and perhaps didn't want to see!

Crabs are the most recognizable crustaceans on the beach, and ghost crab (*Ocypode quadrata*) burrow holes and the trails of footprints between the burrow openings are common on the back beach. The holes are typically ½ to 2 inches in diameter and occur in patches. The shells of other species of crabs can be seen in the wrack line as well.

Another common crustacean you'll encounter is the mole crab (*Emerita* species). If you've stood in the backwash to shallow surf zone, you've probably seen groups of mole crabs appear as the turbulent water exposes their shallow abode. They scrabble down the slope and quickly disappear again beneath the surface of the sand.

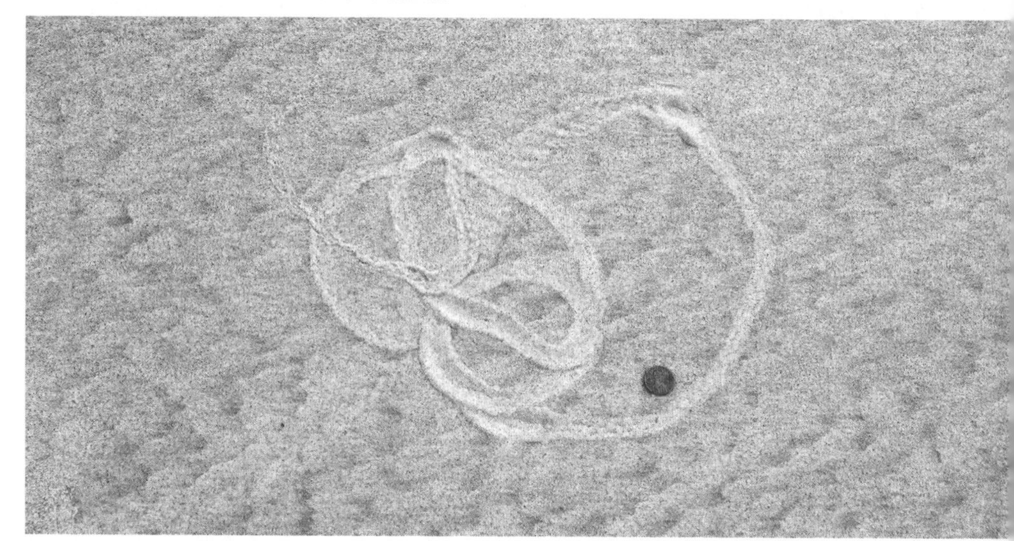

A surface trail pattern on a beach at Ormond Beach, Florida. The looping pattern suggests that the animal was feeding. Penny for scale.

On southeastern U.S. Atlantic beaches the ghost shrimp (*Callianassa* species), another burrowing crustacean, is a common resident of the intertidal zone. Its burrow holes are sometimes abundant, especially on tidal flats. Look for holes with a diameter slightly larger than a soda straw, surrounded by little, dark, cylindrical fecal pellets that look like chocolate sprinkles. As these small animals burrow to considerable depth, they cause water and sand to be extruded from their burrows, which produces sand volcanoes similar to those produced by the physical escape of water and sand. The burrow is more than just a hole. The animal constructs a wall of sand grains to strengthen its burrow in the high-energy environment of the beach, and as

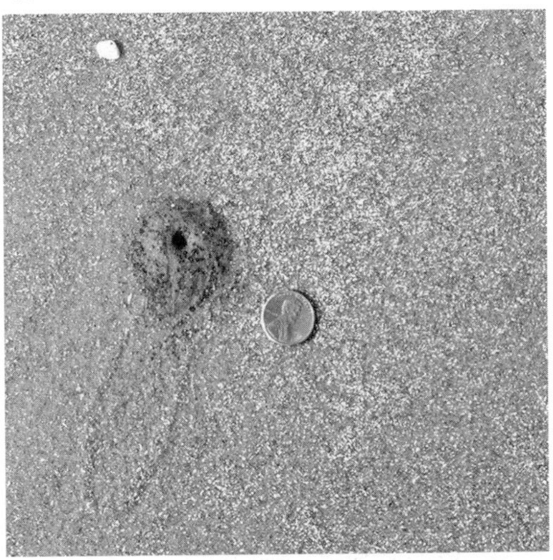

A sand volcano produced by a *Callianassa* species shrimp as it extruded sand and water from its burrow on the tidal flat at St. Simons Island (Georgia). The darker color of the cone is from fecal material and sand of a different color that the shrimp penetrated while burrowing below the surface. Although faint, a flow pattern, produced by the extruded water and sand, can be seen extending from the base of the hole. Penny for scale.

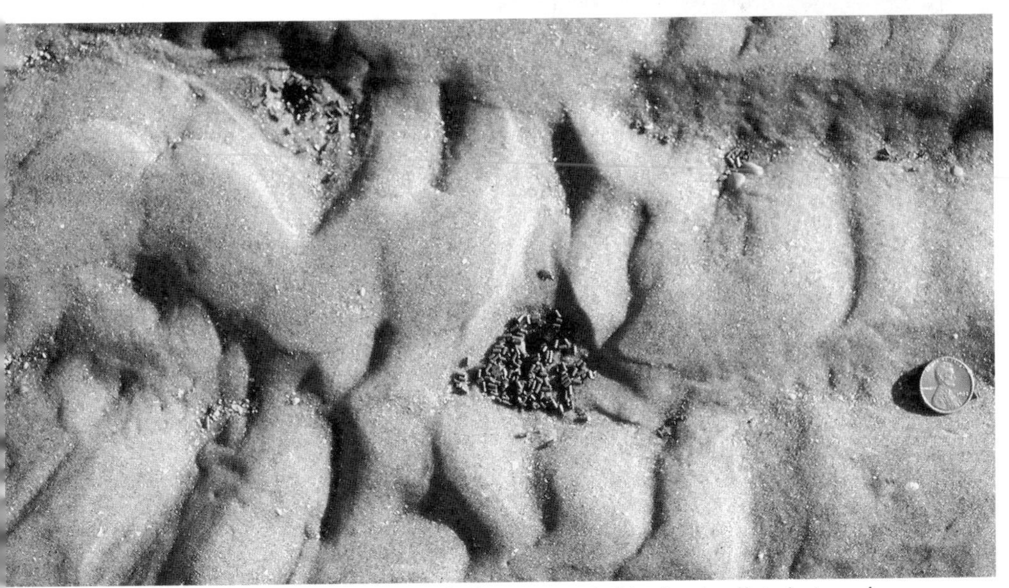

*Callianassa* species shrimp burrow holes surrounded by fecal pellets in troughs of ladderback ripples at Isle of Palms (South Carolina). Penny for scale.

waves or swash scour sand from around the burrow a chimneylike structure may emerge, protruding from the beach.

Another common group of small crustaceans found on the beach are amphipods, sometimes called *sand hoppers* or *sand fleas* (*Orchestia* species), which are mistaken for insects because of their antennae and hopping movement. These critters range from ½ to 1 inch in length and burrow into the sand during the day, usually just above the high tide line. At night they come out to scavenge. Their holes look like nail holes, but these holes usually extend down into the sand for at least a few inches, and often they are not vertical but occur at a slight angle. Sand hopper holes are also more uniform in size than nail holes, and the opening of these holes may be slightly oval with a slight rim. You will see these amphipods during the day as well, and when you approach them they will take shelter in holes, including nail holes, rather than their own burrows. If you're not sure how a hole formed, carefully cut a vertical face through the hole. Burrows will extend well below the surface, whereas air holes will be very shallow, usually less than an inch.

Insects are common on beaches, and some, like sand flies, burrow in search of food or to lay eggs. Sand fly larvae produce recognizable burrow

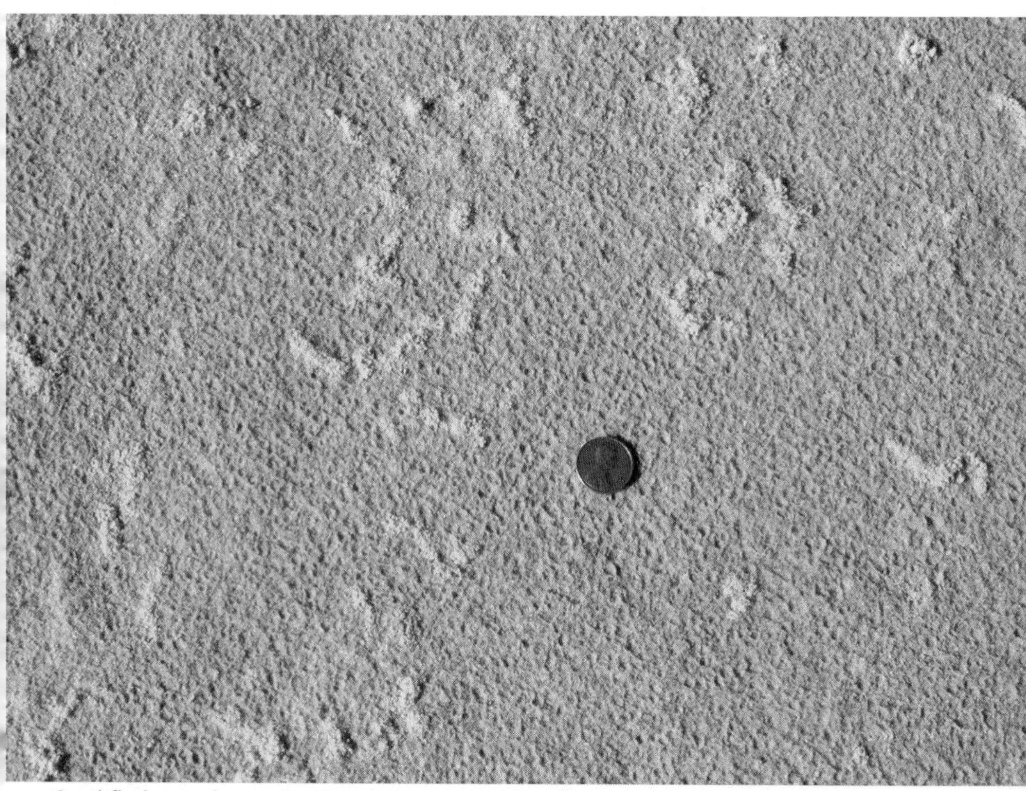

Sand fly larvae burrowing just below the surface formed these irregular, white raised traces. Raindrops produced the pitted surface texture of the beach. Penny for scale.

patterns just below the surface, which appear as short, irregular traces of about ⅛ inch in width on the beach surface.

Although most mollusk shells on the beach originated offshore, there are some common snails and clams that live in the beach. At low tide you may find the trails of small snails that are grazing over the rippled surfaces of the tidal flat. Snails, such as the moon snail (*Polinices duplicatus*), also called shark eye, or the lettered olive (*Oliva sayana*), may work just under the surface and leave convoluted trails or a feature that looks like a small slash in the sand.

The small, colorful coquina clam (*Donax* species) often resides just below the surface, with its siphons extended to produce two small holes in a pit in the surface of the beach. Coquina clams usually gather in groups, so you might see a field of holes that resemble nail holes. If you want to test whether the holes are inorganic nail holes or were dug by coquina clams, carefully excavate around the hole to see if a little clam is present. Breaking waves and swash often expose these beautiful little creatures as well.

You're likely to find traces of other animal groups as well, although they are not as abundant as those mentioned above. Echinoderms, such as starfish

A snail burrowing just below the surface produced this curved slash mark on the beach at Edisto Island (South Carolina). Penny for scale.

and sand dollars, may be present. Sand dollars (*Mellita* species) burrow along just under the sand on beaches with low waves. You can spot them by the elongated surface disturbance of sand, as wide as a sand dollar, and at the end of the disturbance a set of breathing holes over the animal. Scoop your hand under the pattern and you're likely to expose a sand dollar.

### *Beach-Dune Transition*

The area extending from the wrack line into the dunes is one of the best places to look for animal traces. In addition to crabs, a variety of insects are common in this zone, the latter often taking advantage of preexisting holes for shelter. Many insects burrow for protection, to feed, or to lay eggs. In addition to fly larvae mentioned above, beetles, bees, and ant lions each create unique patterns in the sand. Evidence of larger animals is often apparent in the form of footprints and droppings. Even plants leave unique scribe marks as wind blows their stalks and exposed rootlets.

A plant scribe mark produced as the wind moved the seeded head of the grass back and forth over the dune. —Photo by Drew Wilson/*The Virginian-Pilot*

Burrowing beetles produce this unique pattern commonly found in sand dunes. Sometimes birds will systematically peck along the raised burrow in search of the beetle. Nickel for scale.

These depressions are abandoned ant lion pits. They are clever traps in which a wandering insect slides down the slope of the pit and is captured by the ant lion. Plant debris has accumulated in these pits in the dunes along the northern New Jersey shore. Also note the circular scribe marks produced by the tips of windblown grass fronds.

A cluster of horse apples on Sea Island (Georgia) tells of the presence of larger animals on the beach. Although grazing animals tend to destabilize dunes, grass seeds germinating from their fecal matter demonstrate a mechanism of seed dispersal and dune stabilization.

### *The Solution to the Mystery*

So what caused the mystery structure shown in the photograph on page 154 It's the trace of one of the most common denizens of the beach: a human footprint seen in cross section. The foot's pressure disrupted the sand laminae just below the surface, contorting the thin laminae. Although the rising tide and swash erase the surface trace of footprints, the subsurface trace sometimes remains as evidence of human presence on the beach. Human mystery prints can be as fascinating to interpret as the trace of any other organism.

The trace of a snake's journey up the face of a dune. —Photo by Drew Wilson/*The Virginian-Pilot*

# Shells

Almost everybody has a seashell in his or her house sitting on a bookshelf, coffee table, or windowsill. And almost everybody remembers the excitement of finding it (unless you actually bought it at one of those seaside shell stores). Often the lucky shell owner came into possession of his treasure by getting up early and walking the beach just as the sun rose, thereby beating out the competition. On some beaches, especially in Florida, avid shell hunters comb the beaches with flashlights when low tide occurs in the middle of the night. They are the truly dedicated shell collectors!

Most people assume that the shell on the shelf belonged to an organism that recently lived happily on the seafloor until it died and the shell rolled up on the beach, where it briefly resided before the lucky shell hunter picked it up. But a nice, simple sequence of events like that is the exception rather than the rule.

As it turns out, and as we shall explain, seashells on the beach got there through a variety of means, and a careful look at their color, shape, orientation, and species type can reveal all kinds of things about beaches. You can even distinguish an artificial nourished beach from a natural one by its shells.

Some of the questions an alert beachcomber might ask include:

Why are there shells on the beach that belonged to organisms that never lived on the seafloor near the beach but, in fact, lived in the lagoon behind the beach?

Why are there fossil shells, thousands and even millions of years old, mixed in with younger shells on a beach?

Why are some shells colorful while others are all brown or black?

Why is the species makeup of the shells on a beach sometimes very different from week to week and even day to day?

Why are some shells in perfect condition while others are in bad shape?

Why do shells often occur in patches on a beach?

Why are clamshells almost always oriented with the cavity, once occupied by the critter, face down on the beach?

How do shells get into the sand flats behind the dunes that line the beach?

The answers to these questions are just the beginnings of a fascinating and complex story about shells on the beach. After you read this section, you are destined to never look at shells the same way again!

## Recently Alive or Long Dead? Beach Fossils

If there is one single object that people relate to beaches, it is the seashell. Taking advantage of this interest, every beach resort in the world has a seashell shop or two, sometimes dozens, selling all kinds of shells. In East Coast shell shops, the merchandise is usually from some other part of the world, although it is not labeled as such. Typically, shells for sale come from tropical waters where shells are produced in greater abundance and diversity. Cowries from the Philippines are a common type of commercial shell. The beautiful conch shells for sale are usually a species from the Bahama Banks, a shell type that has made it all the way to shell stores in California and Hawaii.

More than one East Coast advertisement hawking new condos has featured a beautiful woman walking along a beach holding a conch to her ear to *hear the sea*. Another variation on this theme is a scene with small children playing on the beach with beautiful conch shells that ostensibly they just found nearby. The problem is that the conchs in both these scenes are from the Bahamas or Caribbean and not the U.S. Atlantic Coast!

As it turns out, nature also seems to have conspired to fool us with its shells on many East Coast beaches. The shells aren't always simply what they seem to be: the remains of organisms that lived and died in the beach or just offshore. One of us found this out the hard way. Orrin Pilkey did his dissertation research on the chemical composition of seashells collected on beaches

Surf clams (*Mercenaria mercenaria*) are common on northern New England beaches because their shells are strong and resist abrasion. Here the wind is sculpting the beach surface, removing finer sand from around the shells. Penny for scale.

One might assume that this intertidal shell hash is an assemblage of shells from animals that were recently living, but it is actually a mix of both modern and fossil shells. For example, the oyster shells are fossil shells eroded from old marsh and estuarine muds. Clams dominate the assemblage, but snail shells are present as well. Note that some of the shell fragments have small holes, evidence of borers like certain species of sponges and mollusks that speed up the breakdown of shells into sand-sized particles. —Photo by Andy Coburn

from Maine to Florida. He based his thesis on the assumption that the shells were from organisms that had recently died. The purpose of the study was to compare the composition of the shells with the temperature of the water in which they lived. If a relationship could be proved, it could then be applied to fossil shells in order to determine the temperatures of ancient oceans.

Orrin's basic assumption was all wrong, however. Many of the shells on East Coast beaches are fossils, thousands and even millions of years old. Their chemical composition may be completely unrelated to the present water temperature. Fortunately for Pilkey, universities don't ask for the return of their PhD diplomas if, a decade later, the thesis turns out to be wrong!

The shells on most East Coast beaches may not appear as spectacular as Bahamian conchs, but their history is often a lot more interesting. Shells arrive on the beach in a variety of ways.

**FROM WITHIN THE BEACH:** On all East Coast beaches, a small fraction of the shells in the sand are derived from organisms that lived and died in the beach. Any shells that live in this environment have to be tough as nails to survive storms and rapid changes in beach shape. One of the most widespread shells found in the intertidal zone is that of the coquina clam (*Donax* species). Coquina shells are also called *butterfly shells*, because when the colorful shells are spread open they look like a pair of small butterfly wings.

**FROM JUST OFFSHORE:** The lower beach and shoreface are home to a great variety of shelled organisms whose shells end up on the beach after the animal dies. Waves force the shells ashore. For example, sand dollars are treasured beach finds that originate from the nearshore environment. Veteran beach walkers notice that different species may arrive on beaches in different seasons because of differences in wave energy. For example, large whelk shells are common after winter storms. Differences in wave energy and shell sources along the shore also account for concentrations of shells in patches or as pavements on the beach. Really big waves that flood the beach and wash over into the dunes or flats behind the beach transport shells inland, so shells often occur beyond the beach.

**FROM OFFSHORE ROCKS:** In Maine, New Hampshire, and Massachusetts, shells on the beach may have been cleaned or scraped off adjacent rock cliffs by crashing storm waves. Barnacles and blue mussels (*Mytilus edulis*) are common organisms that cling to rocks in the intertidal zone, so they are a common constituent of nearby beaches. For example, the sand of Sand Beach in Acadia National Park (Maine) is composed of up to 75 percent shells derived from barnacles, mussels, and sea urchins.

**FROM LAGOONAL DEPOSITS:** Shells that resided in the lagoon behind a barrier island come to the ocean-side beach as the island migrates over them. Such lagoon shells are often the most abundant of barrier island beach shells from New York to northern Florida. Often these shells are stained black. By far the most common species in this category on southern and mid-Atlantic beaches is the oyster (*Crassostrea virginica*), but oyster shells are occasionally found on beaches all the way to Maine; for example, at Popham Beach and Reid State Parks.

**FROM ANCIENT SEDIMENTS AND ROCKS:** The seafloor on the continental shelf is not entirely covered with sand. Many areas have rock outcrops of varying ages that are exposed to waves and currents, which erode the rocks and carry shells to the beaches. These shells may range from hundreds to millions of years old. Beaches in North Carolina, particularly between Cape Fear and Cape Lookout, and beaches south of Cape Canaveral (Florida) may have particularly large fossil concentrations from ancient rocks. Similarly, fossil-bearing beds can be exposed in bluffs or cliffs at the back of a beach and find their way onto the beach as a result of erosion.

**FROM BEACH FILL (ARTIFICIAL BEACHES):** A relatively new source of shells on East Coast beaches is the sediment brought in for beach nourishment projects. Sand for beach nourishment is obtained from many places, including the continental shelf, dredged shipping channels, dredged harbors, inlets, lagoons, and from sand sources on land. The seashells within such sand seldom reflect a beach origin. Today, most often the sand for artificial beaches comes from the continental shelf. Shelf sand is of many origins from many past environments and may be a variety of geologic ages. For example, there are large deposits of sand on the shelf that came from lagoons that existed when the sea level was much lower a few thousand years ago.

Both artificial and natural mixing of shells from a variety of sources means that what you collect on the beach could be from long-dead fossils or from recently living animals. Looking at the list of shell sources, it is clear that many of the shells we pick up are fossils, even some that are in such good

The dark color and sharp angular characteristics of this shell material at Emerald Isle, North Carolina, is typical of sediment pumped from old offshore sediments. These shells darkened due to chemical-reducing conditions while they were buried. The fossil shells were broken in transit through a dredging pipe and during placement. Coarse and sharply angular beach nourishment sediment is often of poorer quality relative to recreational use than the natural sand it replaces. —Photo by Andy Coburn

shape they appear to have lived yesterday. Back in the 1970s, when the complexity of beach shells was first recognized, geologists obtained radiocarbon dates for samples of beach shells by mixing up and grinding together a large number of shells from one area of a beach. These composite dates, although few in number, are very revealing. They indicate that the shell fraction of U.S. Atlantic Coast beaches may get older to the south. Shells from Plum Island (Massachusetts) were on the order of 400 years old, and shells from Assateague Island (Maryland and Virginia) were 1,800 years old. Shells from the beach at Surf City, North Carolina, were 7,800 years old, and three dates from composite shell samples from southeast Florida beaches ranged from 8,000 to 13,000 years before present.

Although these data aren't strong enough to write home about, the dates seem to indicate that the percentage of fossil shells is highest on natural South Florida beaches, and gradually the percentage of fossils decreases to the north. North of Long Island (New York), where the barrier island chain ends, the beaches of New England probably have substantially fewer fossils in the shell fraction than the beaches on barrier islands to the south. Why this is the case is not clear. It is most likely because the southern beaches have derived much more of their sediment from fossil shell material of the continental shelf, while land-derived sediment has diluted the shelf's contribution to beach sediment on northern beaches.

### *Other Fossils and Artifacts*

Shells aren't the only fossils on beaches. Fossilized bones and teeth and tusk fragments have been found, at one time or another, on every East Coast beach. Almost always, these materials have been phosphatized; that is, the original skeletal material has recrystallized or been replaced by phosphate, which gives these materials a black color. Fossil shark teeth, ray teeth, and small fish vertebra are the most common nonshell fossil finds. Occasionally, mammal teeth and bone fragments, and even mastodon tusk fragments, show up. Some of the bone and teeth fragments are from Ice Age (Pleistocene) animals that ranged throughout the forests that once covered the continental shelf when sea level was lower. Not all such animal remains, however, are fossils. A colonial-era cow's horn was found on Brigantine Beach (New Jersey), part of an early garbage dump.

Most of the marine materials, especially the shark teeth, weathered from rock outcrops on the continental shelf and washed ashore. These marine vertebrate fossils are commonly on the order of 15 to 20 million years old; for example, those found at the north end of Edisto Island (South Carolina) and Onslow Beach and Topsail Island (North Carolina). Sometimes the fossils on beaches are rock, such as the worm rocks at Cape Canaveral (Florida) composed of calcium carbonate serpulid worm tubes that look like jumbles of spaghetti.

Human artifacts from long ago and not so long ago also find their way to beaches in the same manner as fossil shells. People occasionally find arrowheads on Coast Guard Beach (Massachusetts). Colonial-era materials have been found at Hilton Head Island, South Carolina, and along New Jersey beaches and the Massachusetts shore. Wagon wheel marks and horse hoof prints have been observed in mud layers on the outer beach at Nauset Light Beach (Massachusetts). Hurricane Hugo uncovered a stash of Civil War relics on Folly Island (South Carolina) where Union troops buried a cache of materials, including boots, belts, weapon parts, and bullets. Nineteenth- and twentieth-century artifacts, particularly shipwrecks, appear on beaches from time to time. For example, two 5-foot diameter tanks from a sunken barge in the Canaveral National Seashore (Florida) periodically appear in the surf zone after storms. Structures left behind by the retreating shore that are now on the beach include the old railroad station platform at Ludlam Island (New Jersey), old gun emplacements off Bogue Banks (North Carolina), and the remains of lighthouses and other structures. At Mantoloking, New Jersey, flat, pointed stones have been washing ashore for decades. The objects, which are 6 to 8 inches long with ½-inch-wide pointed ends, are Italian whetstones from a sunken ship's cargo. When scanning for shells, you never know what other fossils or curiosities may come to light.

## Staining and Rounding: More Than Meets the Eye

Look around you and observe the color of the seashells. If you are on a natural southeastern U.S. beach, one that has not been artificially replenished, a lot of the shells will be brown. Brown is not the original color that a living organism imparted to the shell. This secondary coloration of the shell is caused by inorganic reactions that occurred long after the animal died. Iron oxide, in the form of microscopic crystals of the rust-colored mineral limonite, forms within the minute cavities of shells to color them light brown. The iron possibly comes from the weathering of iron-rich heavy minerals in the sand.

Shells stained brown are found on beaches from Miami to Maine, commonly making up 5 to 30 percent of all the shells on a given beach. There seems to be no particular trend in the abundance of brown shells on beaches along the East Coast. However, since the total shell fraction of beaches north of Cape Hatteras is usually less than 1 percent, brown shells become a minor component of beaches north of the cape. This lack of shells also means that shell color doesn't contribute much to overall beach color.

Brown staining is temporary once a shell has been removed from the beach. For example, shells exposed to weathering on overwash fans that extend landward from a beach seem to lose their brown color in a mere decade or two. Given a bit more time, virtually all colors, natural or secondary, disappear from shells stranded on land, and the shells become a washed-out gray or white color. Many avid shell collectors have displayed their colorful, treasured beach finds along the margin of a flower bed or on the railing of a porch only to find that with time, the color fades away.

Shells stained black are also found in beach sands from Maine to Florida. Black shells are usually somewhat less abundant than their brown counterparts in natural beaches. In mid-Atlantic and southeastern U.S. beaches, the black coloration is commonly found in oyster shells, but other shells can be black as well. The black coloration is caused by the deposition of microscopic crystals of iron sulfide (pyrite) in microscopic pores within a shell. In large crystals, pyrite is gold, but in microscopic crystal form it is usually black.

This grouping of colored shells, even in black and white, shows the color range from natural (right center), to white and gray bleaching (top), and to shells stained brown and black (center and left).

While brown staining occurs in the presence of oxygen in the atmosphere, black staining occurs in the absence of oxygen as a result of burial in mud. Along East Coast beaches, mud occurs behind barrier islands in marshes and lagoons. Landward of mainland beaches, for example, those in New England, it occurs in marshes, bays, estuaries, and ponds. Black shells on natural beaches of barrier islands can be evidence of island migration. That is, the shells arrived at the beach as the barrier island migrated over marsh mud. A further indication that the shells don't really belong on the beach is the fact that many blackened shells are lagoonal species, like oysters, and are not native to open-ocean beaches. Black shells also may be abundant where dredge-and-fill beach construction has mined buried sediment, which contains fossil shells, from offshore.

Another way that shells can be blackened begins with the seaweed often buried in beach sand. This decaying vegetation provides a reducing (oxygen-free) environment that will blacken shells, a process that is more common on New England beaches than elsewhere due to the abundance of seaweed. In general, such buried seaweed is within easy shovel reach of the beach surface if you care to look for black shells.

You can also blacken your own shells. Take some fresh shells and bury them in mud, preferably salt marsh mud. Dig them out of the mud in a month or more and you will find that most of them have blackened. Some shells blacken in a matter of days; some in a matter of weeks; and some, like scallop shells, never seem to blacken in short experiments like this. If you place some of these newly blackened shells on a sidewalk and let them sit in the sun for a few days, the color will disappear from some species but not others. Such simple shell blackening and bleaching experiments provide neat science demonstrations, either for science classes or for the curious beach stroller. Many questions remain about shell staining on beaches.

Another attribute of seashells that reveals something about local beach conditions is the rounding of shell fragments. *Rounding* refers to the angularity of corners and edges of flat, pebble-sized fragments. Rounding occurs as sand grains are tumbled about in the surf zone, wearing away the edges of shells. In a very general way, the higher the average wave size the more rounded the shell fragments. Thus, beach shells at Cape Hatteras (North Carolina), on average, are more highly rounded than those on other East Coast beaches. In contrast, shells on Jekyll Island (Georgia) are hardly rounded at all. Very well-rounded shell fragments, such as those from beaches near Cape Hatteras, can have beautiful displays of shell color, original or secondary, and are sometimes mounted as jewelry.

There are flies in the ointment of interpreting shell rounding, however. Driving on beaches breaks up shells, producing sharp-edged fragments. Beach nourishment introduces sand to the beach that may or may not have rounded shell fragments. When the sand source for an artificial beach is inland, the beach may have no shells at all.

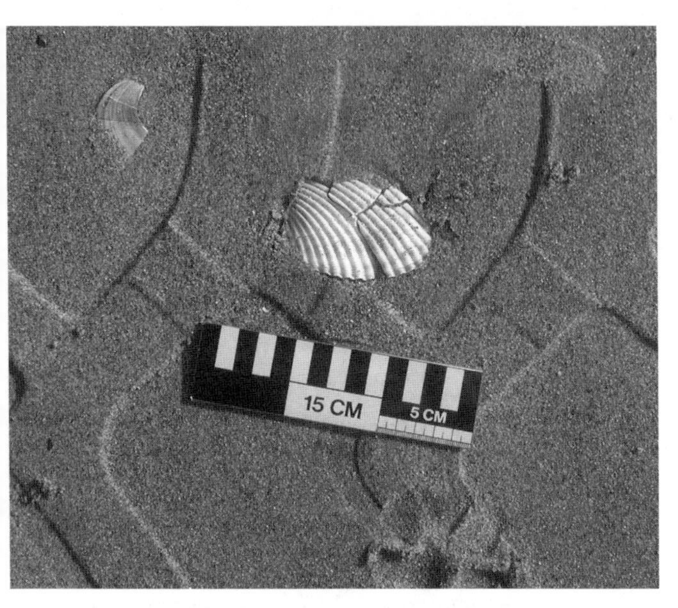


Vehicle traffic on beaches breaks shells and produces angular fragments, as in this tire track on a North Carolina beach. —Photo by Tracy Rice

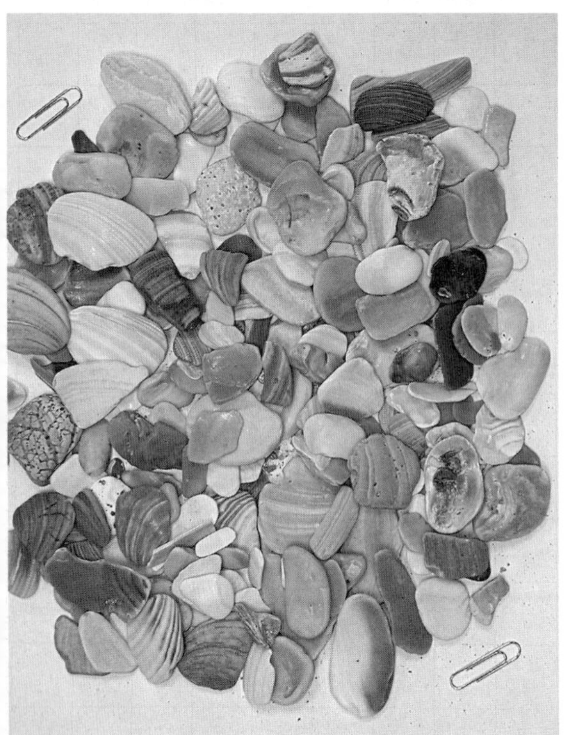

Broken shells range in shape from angular to rounded. This shell hash consists of a high percentage of rounded and polished shells, indicating a long period of abrasion in the surf zone at Buxton on the Outer Banks (North Carolina).

## Preferred Orientation: Hydrodynamics and Aerodynamics

Regardless of color or whether recent or fossil, seashells tell us more about beaches than most beach strollers realize. Take ordinary clamshells, such as arks, cockles, scallops, and oysters, which have good concavity. A shell can reside on the beach in two orientations: concave up, with the living cavity facing up, or concave down. Along most open-ocean beaches, 80 to 90 percent of all shells will be concave down. Out on the continental shelf, however, in deep water beyond the beach, most of the shells are oriented concave up. What gives?

You can demonstrate for yourself why this happens. Wade out to a water depth of 3 or 4 feet, and in between waves drop some clamshells into the water. Even without a face mask, you can see that as they fall through the water column they are always oriented concave up, and that is the position they initially maintain on the seafloor. However, with passing waves, the shells are often turned over so the body cavity faces down. This more stable orientation is easily demonstrated by placing some clamshells on the beach where they will be struck by energetic swash. If the shells are concave up, most will be turned over within a wave or two, and those shells that were concave down to begin with remain oriented that way.

So the secret is that the preferred orientation—the one that is hydrodynamically the most stable—of clamshells in the presence of strong wave swash is living cavity down. Out on the continental shelf, the wave motion on the seafloor isn't great enough to routinely turn the shells over, so most shells rest with their living cavity facing up. Geologists can use such shell orientation in ancient sandstones to distinguish whether the sand was deposited on a beach, in very shallow water, or in deeper water.

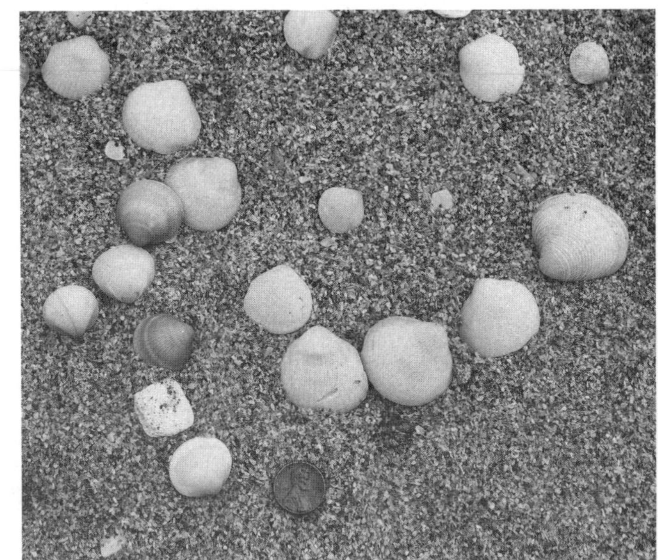

Shells in the typical concave-down orientation on the beach at John D. MacArthur Beach State Park (Florida).

Occasionally, beach strollers may find patches of open-ocean beaches where the shells are oriented "incorrectly." This occurs where an offshore bar of some sort acts as an offshore breakwater and significantly dampens the strength of the waves striking the beach. We've noticed such anomalous shell orientations at Cape May, New Jersey; the south end of Cedar Island (Virginia); and the north end of Kiawah Island (South Carolina). In most cases, the exceptions were on beaches near inlets where an ebb tidal delta had built up to the point that waves were significantly dampened.

Other kinds of shell orientation may be found on beaches. Elongate shells, such as oysters and razor clams, can have a preferred orientation relative to the shoreline, usually with their long axes perpendicular to the shoreline due to orientation by wave swash. This orientation is rarely as obvious as the concave up or down positioning of clamshells.

A much rarer form of preferred orientation can be found on the upper part of very shelly beaches after a long period (a day or two) of relatively high winds blowing from a single direction. Under these circumstances, the large shells, including clams and conch shells, become oriented in an aerodynamically preferred way. The shells aren't picked up by the wind; instead, they are gradually turned to a common orientation as sand is excavated around the shell edges. We saw this phenomenon once on the Shackleford Banks (North Carolina).

Whether through age, color, or orientation, seashells have interesting stories to tell about their time on the earth. And while the shell collector often looks only for the prized whole, naturally colored shell, the curious look at all of the shells, whole or in bits, to decipher the stories of how they journeyed to the beach and what happened along the way.

Shells in the anomalous concave-up orientation on the beach at Cedar Island (Virginia).

# Dunes and Winter Beaches

The following section addresses two aspects of beaches that often are, but shouldn't be, overlooked: the relationship between coastal dunes and beaches, and how a winter beach contrasts with a summer beach.

Many beaches are backed either by a single row of dunes, the foredune, or by a field of dunes of variable width. Such dunes are an integral part of the beach cycle, and they are a critical source of sand that may sustain the beach during storms. Dunes also provide critical habitat, as well as natural storm protection, for landward areas. We introduce dune types, the origin and processes of dunes, and review some of the common physical and organic structures you are likely to encounter in an active dune.

We introduce winter beaches, particularly those seen in New England, to encourage you to visit beaches in the so-called off season. Ice features and structures, which we describe, develop on beaches due to freezing and thawing, and they are spectacular.

## Dunes: Banked Sand

Many shorelines are typified as much by sand dunes as they are by beaches. Sand dunes provide recreational alternatives—from hang gliding to nature walks—to the typical beach activities. The natural processes of dunes contrast sharply with those of adjacent beaches. Sandy beaches are deposits formed by waves, while sand dunes are formed by wind. Grains of sand in dunes tend to be fine and uniform in size, while beach sand contains a wide variety of sand sizes and coarser material, including pebbles and large shells. In spite of these differences in appearance, sand dunes and beaches do not fare well without one another; they are interdependent.

### *Beaches and Dunes*

Consider the role of beaches in the formation of sand dunes. On a typically hot summer day, the sun's rays warm the land much more than the sea. This differential heating is partly related to the fact that the ocean absorbs sunlight over a considerable depth, while beach sand warms largely at the surface. Thus, when we get to the beach, we must run through the hot sand to the cool water or bury our feet into the sand to avoid getting burned. By noon on a hot day, the beach is so much warmer than the sea that a sea breeze forms. This refreshing breeze occurs when heated air over the land rises and is replaced by cooler air drawn in from over the ocean.

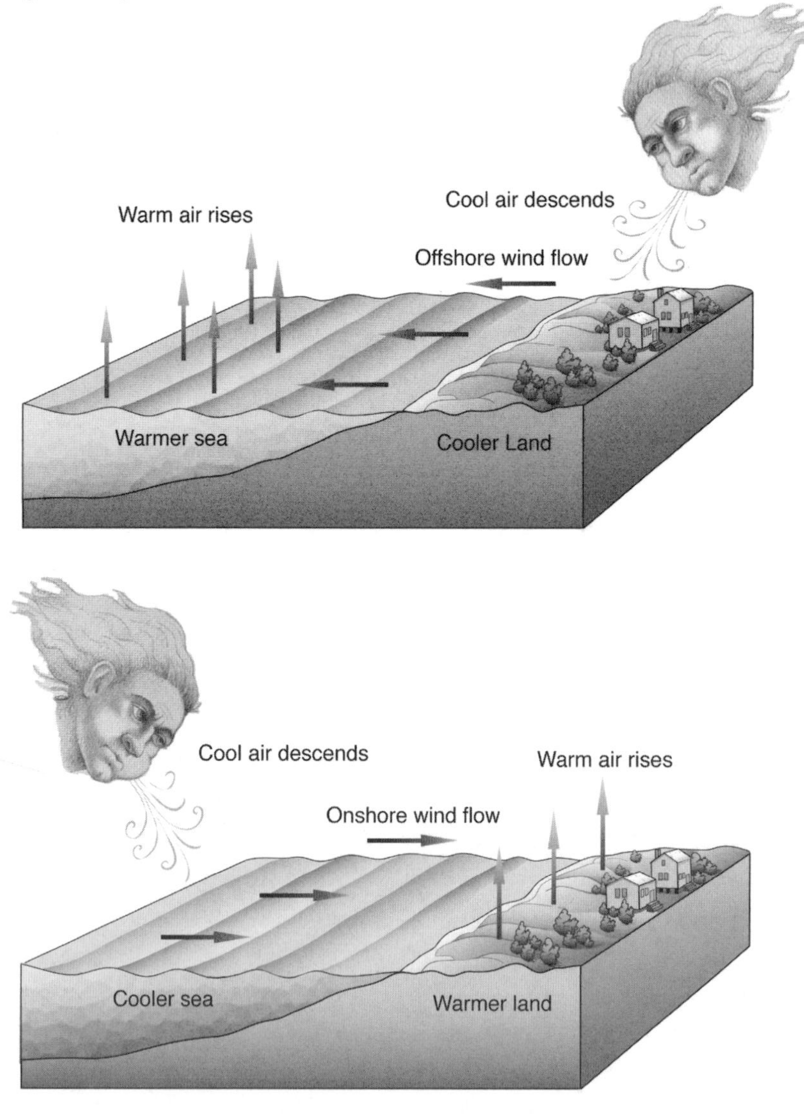


The differential heating of land and sea makes the shore a breezy place. Top: Offshore breeze occurs when warmer air over the ocean rises and air then moves off the land to replace the rising air. Bottom: The familiar onshore sea breeze occurs as the land heats up during the day, causing warm air to rise and cooler surface air to rush in from the sea. —Drawing by Charles Pilkey

As the sea breeze caresses the beach, it picks up finer sand and carries it landward beyond the upper beach into the dunes. Every grain of sand in the dunes once resided on a beach. The sand is transported in three ways: suspension in the air column, saltation (a type of bouncing), and by rolling. The types of transport differ because the wind can briefly suspend very fine sand, while rolling coarse sand but not picking it up. If you can brave a windy day, get down close to the dune surface and observe both the bouncing and rolling grains. Sand arrives at the dune in small amounts at a time and is deposited in very thin, evenly spaced layers, some just a grain or two in thickness (a lamination).

The resulting bedding is visible because different types of grains are segregated and concentrated into layers, depending on the strength of the wind. For example, from Georgia north, heavy minerals often form individual dune layers, representative of periods of strong wind that concentrated the heavy minerals while blowing away the lighter and more abundant quartz grains. Heavy minerals, distinguishable because of their dark color, typically make up around 1 to 5 percent of the sediment of mid-Atlantic dunes. In Florida, the layering in dunes is most often related to the segregation of small seashell fragments. The shell fragments tend to be flat in contrast to the more spherical quartz grains. Such shape differences make the two types of grains

As wind blows sand from around the coarser shell fragments, a lag deposit, or pavement, of shells forms at the back of the beach. Eventually, this armored surface will reduce or cut off the sand supply, slowing dune growth.

Finer-grained, light-colored sand accumulates downwind of pioneer plants and large shells on the Core Banks (North Carolina). Where fine sand is removed from the beach surface, coarser shells and shell fragments are left behind as a lag deposit.

respond differently to the wind. Quartz grains move more readily, leaving the flat shell fragments behind as a lag deposit. When buried, this shell pavement will be a distinct layer. Shell and heavy-mineral layers are often visible in the same dune on southeastern U.S. beaches. In New England, some layers in the dunes are dark-colored, sand-sized rock fragments.

As dunes form, coarser materials, like seashells and larger sand grains, are left behind on the beach as a lag deposit, a coarse armor on top of the beach. Size discrimination by the wind makes dune sands the best-sorted sand deposits in nature. That is, almost all of the grains of sand in a given dune layer are the same size.

When blowing sand grains encounter dune grasses, other plants, or obstructions such as driftwood and other wrack at the back of the beach, the grains drop out of the air due to reduced wind velocity. On a windy day you can see little piles of sand forming around clumps of plants or piles of seaweed and behind logs, timbers, or any other kind of obstruction on the beach; for example, bottles, cans, and beach chairs. This sand-trapping effect of natural wrack lines is a strong argument for leaving them on the beach rather than "cleaning" the beach. Sand accumulates behind, or in the lee, of such objects—in the object's shadow, which is sheltered from the wind. The resulting shadow dune, also known as an *embryonic dune*, may be destined to grow into a mature sand dune; or it may be washed out by the next high spring tide. These initial sand accumulations grow and coalesce into the sand dune line at the back of the beach, which is usually called the *foredune* or *frontal dune*. The foredune is often the largest of the dunes, because the dune itself checks the wind's velocity and creates conditions that are favorable for more sand accumulation.

The biggest Atlantic Coast dunes are the *médanos*, which is Spanish for "coastal sand hill." These high, steep sand hills are usually not vegetated, range in height from tens to several hundreds of feet, and incorporate tremendous volumes of sand. They originate from winds of variable direction and large sand supplies. Examples include several well-know North Carolina locations, including Jockey's Ridge and Run Hill Dune at Nags Head, and Kitty Hawk. All three examples are now park land, after citizen-led efforts prevented the development of some of these dunes. Jockey's Ridge is popular with hikers, hang gliders, and people who climb the dune for the challenge or the view, but the heavy foot traffic is contributing to the dune's changing shape, including its loss of elevation. Jockey's Ridge is mostly barren of vegetation, allowing the wind to continue moving its sand. In contrast, Kill Devil Hills, the site of the Wright Brothers monument, is stabilized by the grassy lawn that the National Park Service maintains.

An embryonic dune forming around pioneer sea oats (*Uniola paniculata*) on the back of a beach on the Outer Banks (North Carolina).

The relationship between beach and dune processes can be seen in its various stages along the beach-dune boundary. Seeds of salt-tolerant plants, blown onto or washed up on the back of the beach by the high tide, germinate to grow into the seaward-most plants on the beach. These pioneer plants become the initial sand trappers that form shadow dunes. These embryonic dunes provide enough elevation above storm water to provide a platform for additional plant growth and more sand trapping. As the plants grow, so do the newly forming dunes. Continued growth of the dunes advances this cycle, providing increased shelter to an expanding list of plants and, eventually, allowing the growth of maritime forest. Any type of dune without protective plant cover is subject to erosion and sand movement. This deflation—the wind's removal of sand from dunes—produces blowouts (bowl-shaped erosional depressions) and U-shaped parabolic dunes that migrate; for example, the dunes east of Provincetown on Cape Cod (Massachusetts).

### *Vegetation and Dunes*

Of the variety of plants that occur in sand dunes, most share three important properties. First, dune plants can tolerate wind and salt spray. Most terrestrial plants are killed by salt or consistent strong winds, so this tolerance is a special adaptation for life near the sea. The plants in the frontal dune are the most tolerant of salt.

A frontal dune ridge, or foredune, at the back of the beach at Seawall Beach (Maine). Typically, the frontal dune is higher than dunes behind it. Some pioneering dune plants also appear to be growing seaward of the dune. Such plants may initiate new dune growth or be killed by the next storm or high spring tide. American beach grass (*Ammophila breviligulata*) tends to form continuous frontal dunes such as the one seen here.

Second, dune plants can survive in dry, desertlike conditions. Highly porous dune sands retain water poorly, so dune plants have adapted strategies for conserving freshwater. For example, after a dry spell, look at the leaves of the frontal dune plants. Some plant leaves curl up to minimize the area through which they lose water by evaporation and transpiration. Other dune plants, like dusty miller (*Artemisia stelleriana*), have hairy leaves, which also reduce evaporation. Some plants, such as the sea rocket (*Cakile edentula*), are succulent, or have the capacity to store water. Yet other plants, like the common dune shrub bayberry (*Myrica* species), have waxy leaves to prevent water loss.

Third, dune plants generally can recover from disturbance by storms. Many dune plants possess extensive horizontal roots called *rhizomes*, from which individual shoots emerge. If part of a dune is eroded or washed away, the rhizomes from what remains can grow back over the disturbed area. In contrast, such recovery is less likely where continuous disturbance due to foot paths or vehicular traffic occurs. Those “Keep Off the Dune” signs should be heeded.

The cutting of maritime forests allowed large parabolic dunes to form in the Province Lands at the Cape Cod National Seashore (Massachusetts). These dunes were formed by strong northwest winds and are advancing toward the camera (view is to the north). The ends of these parabolic dunes are anchored by shrubs, but lacking more vegetative cover the sand advances on the adjacent lake and road. These dunes may have first been activated by colonial-era settlers who cut down the original forest. —Photo courtesy of Jim Allen of the National Park Service

This erosional scarp in a frontal dune has exposed the long roots and rhizomes of dune plants. Such meshes of plant roots help hold sand in place, but the roots also disrupt the subsurface layering and other bed forms. Note the layering of sand is visible because of heavy-mineral horizons, and the dune is growing either on previous beach or overwash sand, the coarser, layered sand in the middle of the scarp face.

Because coastal conditions range from impossible (for example, the lower beach) to difficult (for example, the back-dune area), only a few plant types are adapted to the specific habitats of the back beach and dunes. A single type of plant may dominate such habitats, for example, grasses on the frontal dune; however, careful examination usually reveals a variety of plant types.

From New England to as far south as the Outer Banks of North Carolina, the dominant frontal dune plant is American beach grass (*Ammophila breviligulata*), sometimes called *marram grass*. South of Cape Hatteras (North Carolina), the dominant plant is sea oats (*Uniola paniculata*). These grasses have a common trait; they develop a mass of fine roots that extend to a depth great enough to allow the plants to survive dry periods and to hold their sand foundation in place, and they have rhizomes that permit the plants to send up shoots and spread rapidly across a wide area. The activity of rhizomes is often apparent where a line of new plants starts growing in a nonvegetated

Rows of plants grow seaward across the back of the beach. The stems grow in a line from a single rhizome, or horizontal root, that is slightly below ground.

area. There is a basic difference between the two grasses that is responsible for significant differences in the shape of northern and southern dune fields on the U.S. Atlantic Coast.

The southern plant, sea oats, tends to grow in clusters or clumps; although it does have rhizomes that extend laterally, it is more efficient at growing upward than laterally. The resulting dunes grow around grass colonies rather than in a continuous line. Because of this, frontal dunes along the southern Atlantic Coast tend to consist of a line of individual dunes with numerous gaps—gaps that more readily allow overwash. Storm-surge overwashing is an important mechanism for bringing additional sand and coarser material such as shells into the low areas between dunes.

On the other hand, American beach grass tends to grow in lines parallel to the shore, and dunes form around this grass in long, linear dune lines without gaps. There are, of course, exceptions to this rule. The beaches of Shackleford Banks (North Carolina), where sea oats and frequent dune gaps prevail, and the continuous dune line on Plum Island (Massachusetts), which is vegetated with American beach grass, are good contrasting examples where

Dunes that form where sea oats (*Uniola paniculata*) is the dominant plant tend to be discontinuous, with lows or gaps that often become blowouts or overwash routes where waves carry more sand into the dune system. Also note the stoss (windward) and lee (downwind) faces of windblown sand deposits in the gaps between clumps of sea oats. —Photo by Drew Wilson/*The Virginian-Pilot*

the rule is followed. These grasses illustrate the importance of vegetation in forming dunes, and that preserving dune vegetation is critical to preserving the dunes themselves.

### Active Dunes

If dunes are not vegetated, they remain active, either growing or shrinking in size, and migrating. The fact that sand is being transported to and from dunes is evident on bare surfaces that are covered with wind ripple marks. Wind carries sand up the windward (stoss) side of the dune, depositing it on the face of the dune up to the crest. Sometimes wind carries sand over the crest where it falls on the downwind (lee) face like precipitation. When the crest and lee dune face become oversteepened (an inclination of more than 30 degrees, the natural angle of rest for dry sand), sand avalanches down the steeper downwind face, often flowing in beautiful arcuate fans or lobes. So as sand is eroded and carried from the stoss side of the dune and deposited on the lee face, the dune form moves in the direction of the prevailing wind. In this way dunes migrate and are capable of burying anything in their path.

The sliding, encrusted slip face of a Georgia coastal dune. Once formed, dunes migrate as individual grains are carried over the dune crest onto the lee face and by the avalanching of this face.

Rainfall may give the surface layer of sand some coherence, and it may slip or slide as a crustlike layer down the lee face of the dune. As successive layers of sand accumulate on the sloping dune faces, a unique internal bedding pattern forms; it is called *cross-bedding* and is typical of dunes. Where a natural or artificial vertical cut through a dune exposes its internal structure, the inclined cross-beds, which typically dip anywhere from 5 to 30 degrees, contrast with the more nearly horizontal bedding that one sees in a beach. Sometimes the edges of the inclined beds are exposed on flatter, nearly horizontal surfaces of eroding dunes, producing curved lines across the dune. If the dune face is cohesive and runoff from rainfall is rapid, the face of the dune may be eroded to develop rills or larger gullies.

Active dunes sometimes migrate fast enough to bury trees, roads, or buildings. Good examples of active dunes can be found along much of the Atlantic Coast, including Little Talbot Island State Park (Florida), Cumberland Island (Georgia), northern Edisto Island (South Carolina), Jockey's Ridge and Run Hill Dune (Nags Head, North Carolina), Cedar and Parramore Islands

An unusual rill pattern in the lee face of a sand dune. The wet sand was cohesive enough to allow runoff to erode these rills. —Photo by Drew Wilson/The Virginian-Pilot

(Virginia), Assateague Island National Seashore (Maryland and Virginia), Island Beach State Park (New Jersey), and Monomoy Island and Sandy Neck on Cape Cod (Massachusetts).

### *Plant Succession*

Natural dune settings typically will be stabilized by vegetation, and they may show what botanists call *plant succession*, in which a predictable sequence of plant types occur in a landward direction. Going from the back of the beach into the dunes, one passes into the initial zone of sparse, salt-tolerant, pioneer plants. A dozen or so species of plants are found in this harsh environment. The specific plants vary with the changing climate from Florida to Maine, but some common forms are sea rocket (*Cakile edentula*), Carolina saltwort (*Salsola caroliniana*), sandwort (*Arenaria* species), sea-beach orach (*Atriplex arenaria*), and a couple of species of morning glory (*Ipomoea*). These plants trap sand to form embryonic dunes in which grasses take hold. The grass zone of the frontal dune gives way to shrub thickets. The zone of shrubs grades into maritime forest, which is a climax plant community, the final, mature end product of plant succession.

Geology and botany come together to produce these dynamic habitats. In an area where more sand is added to a beach than is carried away, a series of dune ridges grows seaward over a long time. This “long time” may be on the order of tens, hundreds, or thousands of years. As the shoreline grows seaward, the original frontal dune becomes an inland ridge—on which shrubs and trees may grow—as a new foredune grows seaward of it at the back of the beach.

A back-dune community dominated by American beach grass (*Ammophila breviligulata*), which is not in seed, grading into a shrub community dominated by bayberry (*Myrica pensylvanica*) at Ferry Beach State Park (Maine).

A maritime forest community dominated by pitch pines with beach heather (*Hudsonia tomentosa*) and a lichen (*Cladonia* species) as ground cover at Ferry Beach State Park (Maine).

Palm trees and pines on the beach at Huntington Beach State Park (South Carolina). This out-of-place maritime forest indicates rapid shoreline retreat.

Each dune's history is associated with the sequential development of plant zones, and the changes observed in the plant communities across the dunes reflect a geological history of beach or barrier island growth. Such vegetation zonation can be seen best in undeveloped areas along North Carolina's Highway 12 on the Outer Banks and Florida's Guana River and Little Talbot Island State Parks.

On many barrier islands, however, the plant zonation does not exist due to shoreline erosion over the past few centuries. The zonation has been compressed or destroyed as shorelines have retreated back into the shrubs and forests. Trees now appear next to and on the beach. In New England, the trees may be pitch pine (*Pinus rigida*). Further south, cedars (*Cedrus* species) and live oaks (*Quercus virginiana*) may be prominent. In South Carolina, palmettos (*Sabal* species) may be jammed up against the beach. Walking through the maze of downed trees on the beach at the north end of Jekyll Island (Georgia) or at Huntington Beach State Park (South Carolina) is an adventure. The beach still exists in these places, though the zones of plant succession have eroded away—the beach having retreated over them. However, when the beach position restabilizes, a new plant succession will develop.

A very effective sign to discourage pedestrian traffic on this dune at Cape Henry (Virginia).

Sand dunes are an important habitat for animals as well as plants. We occasionally have observed snakes in sand dunes and commonly see their trails. The most effective signs we have encountered to discourage people from walking over dunes, however, warn about nonexistent snakes at Cape Henry (Virginia) and Waveland, Mississippi.

## *Dunes and Beaches*

Although sand dunes owe the beach for all the sand they receive, they also contribute sand back to the beach. Think of the dunes as sand deposited in the bank, saved up for a rainy (stormy) day when waves will borrow from the bank to maintain the beach. The sand released to the beach by wave erosion accumulates in the nearshore area and causes waves to break sooner than they would otherwise, thus reducing a wave's erosive energy. This natural feedback between the beach and dunes is essential to maintain a healthy beach system, for example, as with the beaches on outer Cape Cod (Massachusetts).

When dunes are removed or seawalls constructed between the beach and dune to protect property, the interaction between beaches and dunes stops. Sand deposited in the bank is no longer available, and the beach narrows during storms. Similarly, people disturb dunes in less conspicuous ways. Dune plants are amazingly resilient, but they do not endure repeated foot or all-terrain-vehicle traffic very well. Trampling through dune grasses breaks their rhizomes and kills plants, resulting in barren dunes that reactivate and form blowouts.

There is another relationship between beaches and dunes that is less widely recognized. This is the poststorm contribution of sand *back* to the beach from the dunes, a very important part of the beach's recovery under the right conditions of strong offshore winds and sand availability.

The final word on beaches is that their very existence depends on their sediment budget. The sandy beaches of the Eastern Seaboard are a dynamic end product of the exchange of sand between each individual beach and adjacent or nearby dunes, sandbars, tidal deltas, inner continental-shelf surface sand, and sand derived from rivers and estuarine sources. Waves, tides, currents, wind, and floods may be the agents of borrowing and lending, but ultimately a healthy sand supply is necessary for a healthy beach.

## Winter Beaches: Frozen Features of the North

From North Carolina to South Florida, beaches are year-round destinations; but from the Outer Banks (North Carolina) to Maine, we associate beach recreation with the summer season, from Memorial Day to Labor Day. Refreshing sea breezes and cool waters make the beach a popular destination to cope with summer's heat. But beaches are still there in winter, and growing numbers of visitors have found that the beach does not lose its charm in a colder season—even in the frozen north. Although the water is not inviting, the beach remains an endless source of interest and discovery, including sights that do not appear in the summer. And in winter there are no crowds of people!

Winter is the time of nor'easters, and, not including hurricane season, beaches from Florida to Maine experience their greatest wave energy from fall through spring. The result is the development of a winter beach profile, which is narrower between the high and low tide lines and has a lot of interesting sedimentary bed forms. Be wise and do not go onto the beach during a storm, but a visit to the poststorm beach will reveal plenty of evidence of a storm's high water level and wave energy. The wrack line often has more coarse debris and debris variety than you find after summer storms, and windblown sand features are common. You may find fresh overwash penetrating the dune line. A lot of foam is generated by storm wave activity, so it's a good time to look for foam features, too.

Occasionally, a nor'easter will bring snow and freezing conditions to beaches from North Carolina to New York. But to see sedimentary structures produced by the freezing of a beach, New England is the best bet. Unique winter features typically require temperatures well below freezing and strong, dry winds. The extreme cold temperatures are required because salty seawater does not freeze at 32 degrees Fahrenheit. Salt prevents water molecules from organizing themselves into ice crystals. At colder temperatures, however, the water comes together to form ice as salt is excluded from the ice crystals. As ice forms, the remaining water is saltier and less prone to freeze. At the shore, however, moving water mixes the excess salt away, and on a very cold day, say below 0 degrees Fahrenheit, considerable freezing of seawater at the shore occurs. Initially, ice crystals form a layer of slush on the water's surface, but with continued freezing, patches of slush begin to form circular patches of floating ice. These patches collide with each other, resulting in raised rims around their edges. These ice features are called *pancake ice*. If the process continues, a sheet of solid ice may form, extending out from the beach.

Frozen swash develops when freezing occurs in the swash zone, and sand is trapped on the frozen bubbles and foam. This patchy sand pattern is produced when melting occurs. Film container for scale.

Shore-fast ice near Lamoine State Park (Maine) forms before an ice foot develops. Nearshore ice forms more easily here because the beach is sheltered in a bay and has low-energy waves. The ice rests on the bottom and remains frozen to the beach when the tide falls.

An ice foot on Old Orchard Beach (Maine). The pile of sand and ice is several feet high and solid as a rock. Ice does not form on this open-ocean beach every year.

On the beach, the first effect of cold weather is frozen wave swash. The bubbles and foam left from an uprush of waves are fragile, with a great deal of surface area exposed. They often freeze into a snowlike deposit, sometimes with bubble imprints. On a windy day, sand can be blown onto this icy pile and become lodged. In the warmth of the afternoon sun, some or all of the frozen swash melts and leaves a curious pile of sand. Swash-ice deposits are not well understood, perhaps because winter beach conditions are not conducive to studying them. Hang out on wintry New England beaches, however, and you can learn for yourself how ice features form.

As pore water in the surface of the beach freezes, a variety of ice-crystal features may form in the sand. These small, delicate patterns are often bladed in form or occur as sets of radiating crystals, almost like the petals of a flower. The largest ice-formed feature on a beach is an ice foot. An ice foot requires prolonged cold to develop because it forms where waves break near the high tide line. The ice foot begins to form when shore-fast ice accumulates in the breaker zone. Eventually, water pushes this ice onto the beach where waves of the next high tide strike it. Waves break onto the frozen pile and cover it with sand. After a few tidal cycles, a pile of mixed sand and ice a yard high and equally wide accumulates. In more northern locales, people

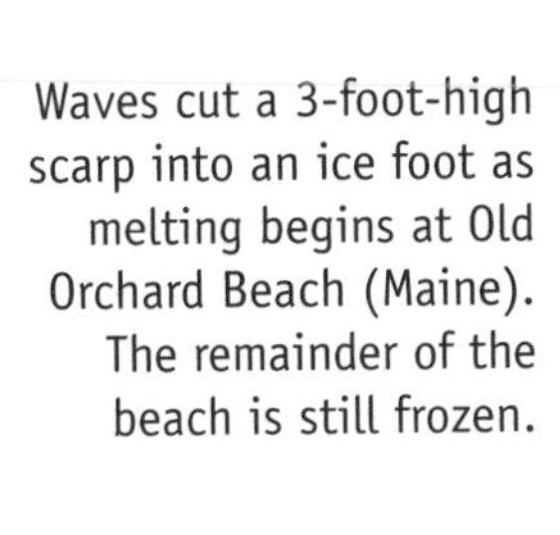

Waves cut a 3-foot-high scarp into an ice foot as melting begins at Old Orchard Beach (Maine). The remainder of the beach is still frozen.

have reported deposits more than twice this height. When warmer weather arrives, the waves and air begin to erode the ice foot, and a small scarp, or bluff, forms on the seaward side of the ice foot. This wall is treacherous to walk over on dark winter days but usually does not last long.

When it is cold enough for an ice foot to form, all water on the beach, including shallow groundwater, is frozen. It is an odd feeling to walk on a beach that feels like concrete underfoot. There are few geological studies of frozen beaches, but it seems likely that during a large storm, a frozen beach will behave as a solid embankment, allowing waves to run across it to the toe of the dune, bluff, or sea cliff at the back of the beach.

On some low-energy beaches in bays sheltered from large waves, extensive tidal flats exist seaward of the beach. When nearshore ice settles onto the tidal flat at low tide, it freezes to the wet sandy and muddy sediment. The following high tide then floats the ice blocks with attached sand and mud, and waves carry the blocks onto the beach or into an adjacent salt marsh. The miniature icebergs eventually melt, leaving a pile of ice-rafted debris. Likewise, ice freezes onto the top of salt marshes and can tear off hunks of peat.

Even people can leave ice-related features on the beach. The weight of a person walking on a wet winter beach compresses the sand and water into a firm imprint. At low tide the imprint freezes and expands slightly as the water turns to ice, creating a footprint in positive relief. Frozen footprints and other ice-formed features are short-lived, however. When summer comes to northern New England (yes, it does get hot in August!), summer visitors will find few traces of ice-formed structures on the beaches, but salt marshes may bear the scars where blocks of ice tore out chunks of peat in winter.

A layer of ice-rafted debris that waves carried from the tidal flat onto Lubec Beach (Maine). The sand and gravel carried in and on the ice will accumulate on the beach when the ice melts. Mitten for scale.

Footprints on the beach rise in positive relief when the water in the prints freezes and expands.

# Saving Beaches or Killing Them?

Beaches are indestructible. They always exist at the boundary between land and sea, no matter what the level of the sea is; how big or frequent the storms are; and whether the beach is in the Arctic, the tropics, or somewhere in between. Beaches are indestructible, *except* when humans get involved.

People love beaches, and it is a fair statement that on the U.S. East Coast beaches are being loved to death. They are lined with buildings, which decade by decade grow larger and higher, ignoring the fact that almost every beach on the East Coast is retreating. The response to the retreat and associated property loss is surprise, dismay, and declarations of natural disaster. When a beach retreats up against the buildings, the fact that beach erosion would not be a problem had the buildings not been built there in the first place is ignored.

Because of human arrogance, people often feel that they should be able to control nature, such as when they try to stop the sea! Seawalls are built to stop beach erosion, but since the walls don't really respond to the root causes of erosion, the beach continues to retreat and eventually disappears. Sand is pumped onto disappearing beaches, killing the beach's critters while taking a big bite out of the federal treasury and not from the homeowners whose homes were built too close to the beach.

It's fair to say that those among us who really believe that the sea can be stopped, or held back along thousands of miles of shoreline, are the fools among us. Nature bats last at the shoreline, and nature keeps dropping hints about her awesome power. Nevertheless, in places like Waveland, Mississippi, where the sea removed houses twice in the thirty-six-year period from Hurricane Camille to Hurricane Katrina, lots were selling for $700,000 or more in the post-Katrina period.

Though not obvious to all the tourists and residents on the beach, humans are responsible for an increasingly large proportion of the beach retreat problem through our production of massive amounts of carbon dioxide. The gas accumulates in the atmosphere, turns the earth into a global greenhouse, and causes ice to melt, seawater to expand, and the level of the sea to rise. A rising sea level causes shorelines to retreat all the more. In fact, by some estimates, in two or three generations society won't have to worry about the shoreline erosion problem on tourist beaches; instead, the focus and financing will be on preserving coastal cities. New Orleans and Katrina was a wake-up call.

So what's more important, the buildings that line the beaches or the beaches themselves? Different strokes for different folks, and there are different answers to that question depending on whom you ask. But if you are a person who loves the beach, the time has come to educate yourself about its future. The following section is a good start in that direction, and the list of references should help you accelerate your beach learning curve. And then let us hear your voice!

## Concrete Beaches: The Suburbanized Shore

A beachless shore on a sandy coast is difficult to imagine, but significant reaches of the Atlantic Coast were nearing that end in the 1970s and 1980s, prior to the widespread construction of nourished, or replenished, beaches. Miami Beach was beachless, and the vistas from the top of extensive seawalls in New Jersey revealed that the beaches were narrow or absent in front of the walls. Seawalls were common along developed shorelines from Maine to Florida, and groin fields were common on barrier islands as communities attempted to hold the shoreline in place.

In keeping with the philosophy of protecting buildings and infrastructure and keeping them from falling into the sea, people have built beach-hardening structures throughout the twentieth century. This philosophy, however, is

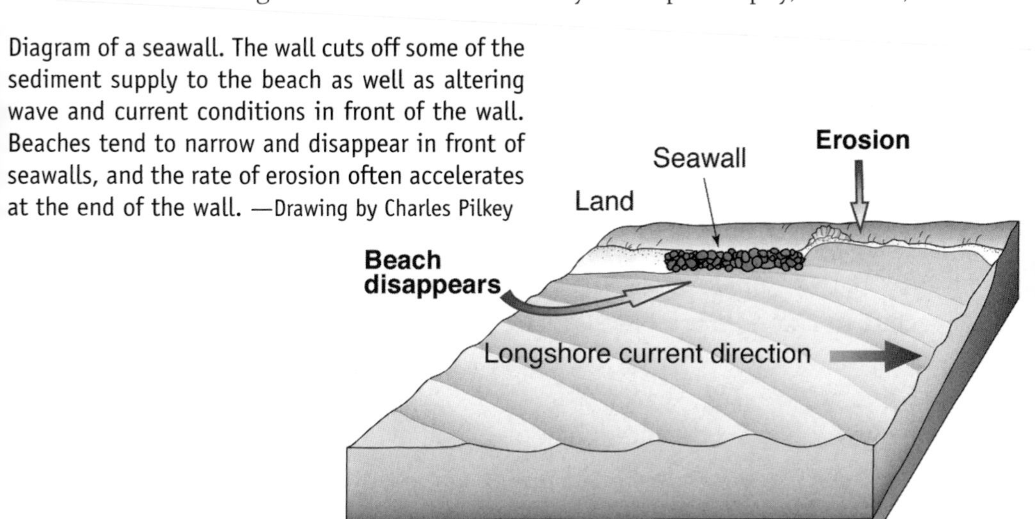


Diagram of a seawall. The wall cuts off some of the sediment supply to the beach as well as altering wave and current conditions in front of the wall. Beaches tend to narrow and disappear in front of seawalls, and the rate of erosion often accelerates at the end of the wall. —Drawing by Charles Pilkey

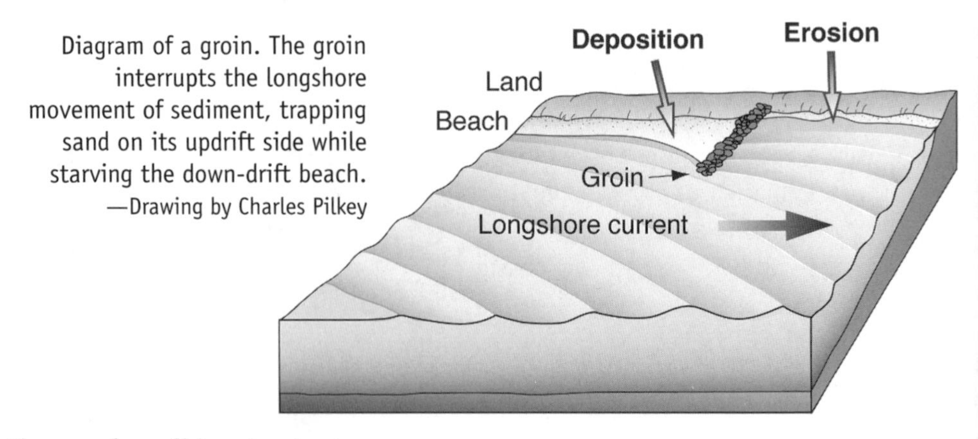


Diagram of a groin. The groin interrupts the longshore movement of sediment, trapping sand on its updrift side while starving the down-drift beach.
—Drawing by Charles Pilkey

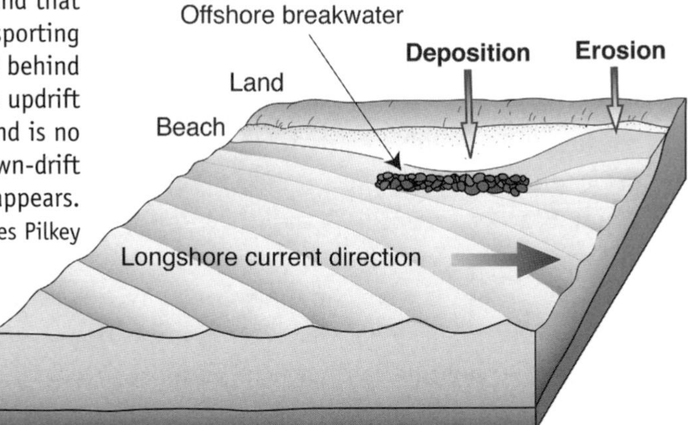


Diagram of an offshore breakwater. The breakwater reduces the energy of incoming waves, thus reducing the longshore current. Sand that longshore currents were transporting along the beach is trapped behind the breakwater, widening the updrift beach. Unfortunately, sand is no longer transported to the down-drift beach, so it narrows or disappears.
—Drawing by Charles Pilkey

counter to one that actually conserves beaches. These structures include-seawalls of various types, groins, and breakwaters. The political and economic momentum for building such structures has usually followed extensive property losses from major storms. Sometimes the loss of life has been the impetus. The 1900 hurricane that struck Galveston, Texas, caused the greatest loss of life (estimated at 8,000) associated with a natural disaster in the United States, and the Galveston seawall was constructed in response to that storm.

The Ash Wednesday nor'easter that swept the U.S. Atlantic Coast in March 1962 caused significant damage to beachfront communities between Massachusetts and northern Florida. That storm prompted construction of new seawalls and improvements to existing walls. More importantly, the loss of beaches in New Jersey, which was blamed on the storm but was mostly due to seawalls, became the focal point of legislation authorizing the U.S. Army

The LBJ seawall on Jekyll Island (Georgia). There is no dry beach in front of the riprap wall at high tide. The sand in back of the wall was thrown over it by waves. The wall has contributed to beach loss.

Corps of Engineers to begin the national beach nourishment program. We discuss beach nourishment in-depth in the following section.

After Hurricane Dora in 1964, Vice President Lyndon Johnson flew over the Georgia coast to review the destruction, and Congress appropriated money to build seawalls on Jekyll and St. Simons Islands. These structures are still referred to as the "LBJ wall"; sometimes Joe Kelley (one of the authors) adds an expletive in front of the name because he fell off the wall in 1980 and broke his shoulder. (Caution your children about climbing on seawalls!) As a result of the construction of the LBJ wall, and walls and groins that followed on Sea Island and Tybee Island, Georgia has armored much of its developed shoreline. Fortunately, most of the Georgia islands remain undeveloped.

Prior to the extensive beach replenishment project in the late 1970s, most of Miami Beach, Florida, was beachless at high tide, mostly as a result of seawall construction seaward of the high tide line. In the 1970s, hotels competed with one another to have the seaward-most walls! The seawall structures, along with groins, quickly strangled the beach, the main attraction for the tourist economy.

The widespread recognition that shore hardening contributes to beach loss led Maine, North Carolina, Rhode Island, South Carolina, and Texas to pass legislation forbidding new shore-hardening structures. In the last several years, however, the emphasis on protecting buildings rather than beaches has tested even these regulations. Management policies, especially zoning, are all deeply entwined with state politics. Florida probably remains the easiest state in which to get permits to build seawalls. Maine is the most restrictive. In North Carolina, only temporary sandbag seawalls are allowed, but "temporary," as politically defined, sometimes becomes permanent.

### *Why Hard Stabilization is Bad for Beaches*

Seawalls, a family of structures including walls, revetments, and bulkheads placed on or at the back of a beach and parallel to the shoreline, create problems for beaches in a number of ways. When placed on a dry, recreational beach, seawalls cause it to narrow. Where beaches are eroding, this initial loss of precious public land can be critical, because the public loses a recreational facility and the community loses the economic draw of the beach. If a seawall is built at the back of an eroding beach, the shoreline continues to move back because the seawall does not address the many causes of erosion. As the beach retreats up against the wall it narrows and narrows until it finally disappears. Seawalls also cut the beach off from its sand supply, which in part comes from the back of the beach. Storms that would normally erode this sand from the back of the beach and place it on the lower beach to protect it

Offshore breakwater structures at Miami Beach, Florida, hold wider beaches on their updrift sides at the expense of down-drift beaches.

cannot do so. This sediment is lost to both the beach in front of the wall as well as down-drift beaches, increasing their erosion rates.

Wall-induced beach loss may take decades to damage a beach to the point where people recognize that the wall is the problem. As a result, the democratic political system has problems responding to beach damage caused by seawalls. That is, if it takes decades for the beach to disappear, we can worry about it later or the next generation can tackle the problem. Beach narrowing happens in front of any stationary structure, so concrete seawalls and sandbag seawalls are no different. Both destroy beaches.

Examples of East Coast beaches damaged by seawalls are numerous. In addition to seawalls mentioned above, other seawalls occur at Rockaway Beach (New York); Cape May, New Jersey; Sandbridge Beach (Virginia); northern Carolina Beach, North Carolina; Myrtle Beach, Folly Beach (the Holiday Inn wall), and south Debidue Beach in South Carolina; Jekyll Island and Sea Island (Georgia); and numerous Florida beaches, including those at Jacksonville Beach, Daytona Beach, and Palm Beach.

Groins, walls built perpendicular to the shore and designed to trap sand in the longshore current, create problems in two ways. First, they trap sand on the updrift side of the structure, starving the down-drift beach. Second, groins cause sand loss from the down-drift beach by directing offshore currents, which carry sand seaward and perpendicular to the beach along the groin. East Coast examples of groin fields (multiple groins) include those found in Deal, New Jersey; North Shores at Cape Henlopen (Delaware); and on Pawleys Island (South Carolina), but singular groins occur all along the East Coast. Offshore breakwaters, built parallel to the shoreline, are also structures designed to interrupt longshore currents. They prevent waves from striking the shore and cause sand to be deposited between the breakwater and the beach. This wave-shadow effect widens the beach behind the breakwater but cuts the down-drift beach off from its sediment supply, contributing to beach erosion. As well as being obstacles, breakwaters are also known to be hazardous to swimmers because they change current patterns. Offshore breakwaters are less common than groins, but they are found on the Connecticut shore; on North and South Beaches at Miami Beach, Florida; and at Winthrop Beach in Massachusetts (a series of five offshore breakwaters known as the "five sisters" and visible from airplanes approaching Logan Airport).

### *Sea Bright, New Jersey*

Sea Bright, New Jersey, is one of numerous communities that came to rely on seawalls for protection in the twentieth century, and its shoreline became

The narrow to absent beach in front of the Sea Bright, New Jersey, seawall prior to the 1999 beach nourishment project. Note the dilapidated groins in the surf zone just below the New York City skyline.

an internationally recognized example of hard stabilization as a response to shoreline erosion. Buildings were saved, but at a high price: the complete loss of the natural beach. Today, as you drive through Sea Bright, your view of the sea is blocked by a high seawall. If you climb the wall, you may be surprised to see a wide beach on the ocean side of the wall, causing one to wonder why the wall is needed. But the beach you are viewing was part of a major beach replenishment project started along the northern New Jersey shore in 1999, the most costly artificial beach project in American history (more than $10 million per mile). Prior to that time the sea was literally knocking on the front door of Sea Bright and had been for decades. Even a moderate 20-mile-per-hour wind blowing off the sea could cause waves to throw water over the wall, a wall where the damage estimate from a 1984 storm ($82 million) was approximately equal to the value of the property the wall was protecting. How did a beach community come to be in such a precarious position?

In the mid-1800s the Sea Bright area was a narrow, undeveloped spit with only a few fishermen's huts, but by 1869 the first permanent houses were built there. Photographs indicate that by 1877 houses lined the shore, close to the high tide line and without dune protection. Ten years later there were wooden bulkheads in front of some of the houses, and in 1898 a rock-rubble wall was constructed at the north end of the settlement. These walls are early indications that property was threatened by the sea, either by erosion or overwash during storms. By 1931 the rubble wall had dimensions similar to the present 12-foot-high seawall, but there was still a beach fronting the structure. Groins were added in front of what is now the downtown. But beach loss continued, until only the wall remained as the last line of defense.

The probability of continuing property losses, and the disaster potential when the next great 1962-level storm struck, led to the end-of-the-twentieth-century beach replenishment project, a continuation of the never-ending saga of developing shorelines.

In spite of the fact that the overall cost of the replenishment project exceeded $200 million in public funding, access to the new beach is extremely limited. Sea Bright, at the end of the nineteenth century, had a railroad platform within yards of the beach where New Yorkers arrived by the thousands, straight from the streets of Manhattan. Today, due to limited beach access—for example, limited seawall walkovers and restricted parking—few from New York, or anywhere else for that matter, can get to the beach in spite of the fact that federal taxpayers paid for it. A motel in Sea Bright next to the beach even boasts of having a private beach. At some locations one can stand on the wall, which is trespassing, and view the skyline of New York and note that only a few dozen people are using the beach.

### *Obstructions on the Beach*

If you are visiting an urbanized shoreline, look around the beach for old or new engineered structures. Unfortunately, on many urban beaches old and abandoned shoreline structures can be a real hazard to swimmers. In locations as diverse as Jacksonville Beach, Florida; Myrtle Beach, South Carolina; Kitty Hawk, North Carolina; and Asbury Park, New Jersey, chunks of concrete from destroyed buildings or boulders from storm-destroyed seawalls can be seen on the beach. On many eroding beaches, bits of pipe, electrical wiring, fragments of swimming pools and septic tanks show up on beaches, especially after storms. Walls at the back of nourished beaches may be covered with sand, such as at Miami Beach, Florida, and Jekyll Island (Georgia). Or the original wall may still be apparent, such as those along northern New Jersey beaches; at Cape May, New Jersey; and at Coney Island and Rockaway Beach in New York. Where beach nourishment hasn't taken place in front of a seawall, the beach is usually absent, for example, at Palm Beach, Florida, or narrow, such as portions of Jekyll and St. Simons Islands (Georgia). Groins, common on beaches fronting many older communities, will be obvious as they block your

A groin field at Deal, New Jersey, blocks the longshore flow of beach sand to down-drift beaches. The seawalls between groins prevent new sand from being added to the back of the beach.

stroll along the beach, such as at Deal, New Jersey. Shore-hardening structures are also obstacles to wildlife, including birds and turtles.

Although the beach may be narrow in front of a seawall or down-drift of a groin, you will still be able to find the sedimentary structures we described in earlier chapters. In front of the seawall you may find only wet-beach features, and because of wave reflection from the wall or wave refraction (bending) around groins, you may find some interesting interference ripples or ladder-back ripples. Current activity adjacent to groins and walls may produce more scour structures, and shell lag deposits are often common near groins. Groins provide visible evidence of sand transport by longshore currents. Sand accumulation, called a *fillet*, occurs on the up-current side of a groin, and erosive narrowing of a beach occurs on the down-current side of a groin. We've also found good rill marks on steepened beaches associated with the groins on Pawleys Island (South Carolina).

## Artificial Beaches and Dunes: Virtual Reality?

Change is the rule at the shoreline; but change in America's beaches, especially Atlantic Coast beaches, has taken a turn by which humans, rather than nature, are the force of change. As beach purists, we hold that many of these changes are for the worse. We love the beach, the critters it holds and

A wooden wall at back of Debidue Beach (South Carolina) is an obstruction to turtle nesting, even though the sign notes that this is a nesting area.

nurtures, and the complex mechanisms that make it work. We're intrigued by the physical processes that one can observe and understand if one learns how to read the beach. So we are biased in favor of natural beaches and believe there is no substitute.

When clean, well-sorted white to yellowish brown crystalline sand beaches are replaced by gray sediment mixes; when an entire nearshore ecosystem is decimated by the dredging and pumping of sand onto the beach; or when natural dunes are flattened to build even more property that will be endangered by coastal processes, we are offended. Reality, however, tells us that the trend of replacing eroding beaches artificially will continue, and within a couple of decades it is likely that most East Coast beaches will be nourished, primarily by material dredged from offshore.

Although we accept that such change is inevitable, we don't like it, and that is why the following discussion emphasizes the importance of understanding the problems associated with beach nourishment. Perhaps such knowledge will motivate more people to encourage conservation of the public's beach over the preservation of property built too close to the beach. In the long-term, removing or moving buildings back as the beach migrates is the only way to conserve natural beaches.

This North Carolina beach, part of the southern Outer Banks known as the Crystal Coast, is no longer crystalline because the nourishment sand from offshore is full of black shell debris and darkly stained grains. The original beach was the same light color as the dune in this photo. —Photo by Andy Coburn

At the end of World War II virtually all U.S. beaches were as natural as the leaves on the trees and the fish in the sea, except those shorelines that were heavily fortified with seawalls. Then came the 1962 Ash Wednesday Storm, the greatest East Coast storm of the twentieth century in terms of size and intensity. The storm's narrowing of the already narrow beaches in front of seawalls in New Jersey ultimately proved to be its most lasting effect, setting in motion activity that has altered beaches up and down the coast.

As a result of this storm, Congress authorized the U.S. Army Corps of Engineers for the first time to dredge sand and pump it onto the beaches of New Jersey, a move that proved to be the beginning of the national beach nourishment program. By 2004, over $2.5 billion (2003 equivalent) had been spent on more than 125 East Coast beaches, from the Florida Keys to southern Maine.

Beach nourishment appeared to be a solution in which we could have our cake and eat it too. Having learned that hard structures such as seawalls were detrimental to beaches, beach nourishment seemed to simultaneously protect beaches during storms and maintain the recreational draw of the coast. Beach nourishment became the solution of choice to solve the erosion problem and save thousands of beachfront buildings. Artificial beaches have

The 1962 Ash Wednesday Storm damaged buildings along much of the Atlantic Seaboard, particularly along the heavily developed New Jersey shore. The impact of this storm led to the beginning of the national beach nourishment program.

provided the basis for a boom in the construction of buildings along our shorelines, but the nourishment solution has not been a boon to the quality of America's beaches, especially on the Atlantic Coast.

So what is beach nourishment, also called *beach replenishment* or *beach dredge-and-fill*? It's a process in which sand is pumped by dredges from offshore or trucked from inland areas. By definition, replenishment sand is from somewhere else, sand that is new to the beach. Usually, sand for beach replenishment is obtained from the continental shelf, nearby inlets, lagoons and bays, or from inland sand pits. The basic problem with replenishment is that the new sand is not usually beach sand. Beach sand is unique because the beach and the processes that form it are unique. The processes that formed glacial outwash sands, river sands, or tidal-delta sands are not the same as those on the beach, and the resulting sands differ in size, shape, sorting, and composition. These differences are a problem in terms of aesthetics, erodibility, and what kinds of plant and animal life this new sand can support.

Sand being pumped onto a beach from a nearby tidal inlet. New sediment rarely matches the original beach sand in terms of texture and composition. Note the gulls gathering to feast on dead and dying beach critters.

If you visit East Coast beaches, it is increasingly likely that you will be walking into the surf zone atop an artificial beach. Good examples of major nourished beaches to visit include those at Miami Beach, Delray Beach, and Jacksonville Beach, Florida; Tybee and Jekyll Islands (Georgia); Hilton Head Island, Folly Beach, and Myrtle Beach in South Carolina; Carolina Beach and Wrightsville Beach, North Carolina; Virginia Beach, Virginia; Ocean City, Maryland; Rehoboth Beach, Delaware; Cape May, Ocean City, Atlantic City, and Sea Bright, New Jersey; Coney Island and Jones Beach State Park in New York; Gulf Beach (Connecticut); Revere Beach (Massachusetts); Hampton Beach, New Hampshire; and Wells Beach and Camp Ellis, Maine.

A small beach nourishment project might be 50,000 cubic yards of sand placed along a ½-mile stretch of beach, such as several on Topsail Island (North Carolina). A good-sized dump-truck load of sand might be 10 cubic yards, so 50,000 cubic yards is roughly 5,000 dump-truck loads. A large project could involve 10 million cubic yards of sand along 10 miles of beach, as on Miami Beach. Replenishment projects for the Atlantic and Gulf Coasts average about 2 miles in length, but some, like the phased project from Sandy Hook to Manasquan Inlet in New Jersey, are 20 miles in length. The number of nourishment projects has increased in recent years, and many beaches have had multiple nourishments.

A typical midsized nourished beach, for example, Wrightsville Beach, North Carolina, costs between $1 and $3 million per mile of shoreline, but costs can range as high as $10 million per mile for large beaches, such as the northern New Jersey beach from Asbury Park to Sea Bright. That dredge-and-fill project cost in excess of $220 million for a 21-mile beach. The national average cost for a cubic yard of dredged sand is about $5, but costs can range between $2 to $12 per cubic yard. Sand obtained by dump truck from the mainland is usually more costly than dredged sand. The higher the waves in a given area, though, the higher the cost of dredging. The large dredges capable of withstanding high waves are more costly to operate. All things being equal, it is cheaper to dredge offshore sand in Florida than in North Carolina or Maine.

Beach nourishment is not a onetime solution to a beach's erosion problem, and the expected life of a nourished beach is difficult to predict. Beaches at Miami Beach, Florida, have persisted for more than two decades, but this is by far the longest life span of any nourished East Coast beach. Nourished beaches always erode at a rate faster than the natural beaches that preceded them, and nearly all artificial beaches experience local erosion problems, referred to as "erosion hot spots," that require additional nourishment.

A beach scarp near 40th Street at Ocean City, Maryland, in 1989. Newly nourished beaches, like this one, typically exhibit scarping as waves and storms remove the beach fill.

Summary of beach nourishment projects (location, frequency, volume, cost) along the U.S. Atlantic Coast up to 2003. That beaches are repeatedly nourished suggests that this solution requires a long-term commitment for communities to finance and that quality sand supplies will remain available, both of which are potential obstacles to maintaining beaches. Most well-known beaches are no longer in their original natural state. The exceptions are most of the national seashore beaches and most state park beaches. (Source: Program for the Study of Developed Shorelines, Western Carolina University, http://psds.wcu.edu/)

| Beach | Times Nourished | Years |
|---|---|---|
| MAINE | | |
| Pine Point Harbor, Scarborough River | 2 | 1956 |
| Camp Ellis | 8 | 1969–1996 |
| Gooch's Beach, Kennebunkport | 2 | 1985 |
| NEW HAMPSHIRE | | |
| Wallis Sands State Park | 2 | 1963–1972 |
| Hampton Beach | 5 | 1935–1972 |
| MASSACHUSETTS | | |
| Salisbury Beach, Salisbury | 2 | 1953–1957 |
| Plum Island, Newburyport and Newbury | 3 | 1953–1987 |
| Revere Beach, Revere | 2 | 1954–1992 |
| Winthrop Beach, Winthrop | 2 | 1956–1959 |
| Quincy Shore Beach (Wollaston Beach), Quincy | 3 | 1948–1996 |
| Wessagussett Beach, Weymouth | 2 | 1959–1969 |
| Dead Neck, Osterville | 7 | 1953–1985 |
| East Beach, Clark's Point, New Bedford | 3 | 1956–1959 |
| West Beach, Clark's Point, New Bedford | 2 | 1959–1980 |
| RHODE ISLAND | | |
| Town Beach, Westerly | 4 | 1988–1993 |
| CONNECTICUT | | |
| White Sands Beach, Old Lyme | 2 | 1957–1967 |
| Prospect Beach, West Haven | 3 | 1957–1973 |
| Woodmont Shore, Milford | 3 | 1959–1964 |
| Gulf Beach, Milford | 2 | 1957–1966 |
| Sherwood Island State Park, Westport | 2 | 1957–1983 |
| NEW YORK | | |
| Westhampton Beach | 7 | 1962–2001 |
| Moriches Inlet | 4 | 1966–1978 |
| Great South Beach | 5 | 1962–1991 |
| Great Gunn Beach | 2 | 1969–1995 |
| Oak Beach, Gilgo Beach, Cedar Beach | 2 | 1946–1959 |
| Hempstead Beach | 3 | 1990–1995 |
| Gilgo Beach | 11 | 1960–2002 |
| Jones Beach | 5 | 1927–1990 |
| Rockaway Beach | 19 | 1926–1996 |
| Coney Island | 5 | 1923–1995 |

| Beach | Times Nourished | Years |
|---|---|---|
| NEW JERSEY | | |
| Sandy Hook | 7 | 1975–1990 |
| Sandy Hook/Sea Bright | 2 | 1999–2002 |
| Sandy Hook/Deal | 2 | 1995–1996 |
| Sea Bright/Ocean Township | 1 | 1999 |
| Asbury/Manasquan | 1 | 1999 |
| Avon-by-the-Sea | 4 | 1950–1982 |
| Sea Girt | 2 | 1962–1966 |
| Berkeley Township | 2 | 1962–1968 |
| Barnegat Light | 3 | 1963–1991 |
| Long Beach Island | 3 | 1963–1979 |
| Harvey Cedars | 5 | 1962–1992 |
| Ship Bottom | 2 | 1956–1963 |
| Union Township | 1 | 1966 |
| Long Beach | 3 | 1959–1963 |
| Brigantine | 3 | 1962–1996 |
| Brigantine Island | 1 | 1999 |
| Atlantic City | 10 | 1936–1986 |
| Ocean City | 24 | 1952–2000 |
| Upper Township | 5 | 1966–1992 |
| Strathmere | 3 | 1982–1984 |
| Sea Isle City | 5 | 1965–1987 |
| Avalon | 4 | 1987–1993 |
| North Wildwood | 2 | 1966–1989 |
| Wildwood | 2 | 1963–1991 |
| Lower Township | 2 | 1969–1986 |
| Cape May | 13 | 1962–1999 |
| Cape May Point | 2 | 1992–1999 |
| Cape May Point State Park | 2 | 1986–1992 |
| DELAWARE | | |
| North Shores | 2 | 1962–1998 |
| Rehoboth Beach | 2 | 1962–1998 |
| Dewey Beach | 6 | 1962–1994 |
| Indian Beach | 2 | 1962 |
| North Indian River Inlet | 13 | 1961–1992 |
| Bethany Beach | 6 | 1961–1992 |
| South Bethany | 4 | 1989–1998 |
| Fenwick Island | 7 | 1962–1994 |
| MARYLAND | | |
| Ocean City | 6 | 1963–1994 |
| Assateague Island | 2 | 2001–2003 |
| VIRGINIA | | |
| Virginia Beach | 48 | 1951–2002 |

| Beach | Times Nourished | Years |
|---|---|---|
| NORTH CAROLINA | | |
| Pea Island | 7 | 1990–2002 |
| Hatteras Island | 6 | 1974–1992 |
| Cape Hatteras | 3 | 1966–1973 |
| Ocracoke Island | 5 | 1986–1995 |
| Fort Macon/Atlantic Beach | 5 | 1973–2002 |
| Emerald Isle | 12 | 1984–2003 |
| Topsail Island | 6 | 1982–2002 |
| Figure Eight Island | 7 | 1979–2003 |
| Wrightsville Beach | 20 | 1939–1998 |
| Masonboro Island | 2 | 1986–1994 |
| Carolina Beach | 28 | 1955–2001 |
| Kure Beach | 2 | 1998–2001 |
| Bald Head Island | 3 | 1992–2001 |
| Long Beach | 3 | 1986–1993 |
| Oak Island | 2 | 2001 |
| Holden Beach | 11 | 1971–2001 |
| Ocean Isle | 8 | 1974–2001 |
| SOUTH CAROLINA | | |
| Myrtle Beach | 3 | 1987–1997 |
| Pawleys Island | 3 | 1988–1999 |
| Debidue Beach | 2 | 1990–1998 |
| Edisto Beach | 2 | 1954–1995 |
| Isle of Palms | 2 | 1983–1984 |
| Folly Beach | 3 | 1982–1993 |
| Seabrook Island | 3 | 1982–1990 |
| Hunting Island | 5 | 1968–1991 |
| Hilton Head Island | 5 | 1973–1997 |
| GEORGIA | | |
| Tybee Island | 5 | 1976–2000 |
| Sea Island | 4 | 1964–1990 |
| FLORIDA | | |
| Fernandina Beach | 5 | 1979–1996 |
| Amelia Island | 4 | 1983–1994 |
| Mayport/Kathryn Abbey Hanna Park | 7 | 1966–1994 |
| Jacksonville Beach | 8 | 1963–1996 |
| Anastasia State Park/St. Augustine | 5 | 1963–2000 |
| Ponce Inlet (north beach) | 6 | 1974–1996 |
| Cape Canaveral/Cocoa Beach | 5 | 1966–1995 |
| Patrick Air Force Base | 2 | 1985–1996 |
| Indialantic/Melbourne Beach | 2 | 1981–1985 |
| Sebastian Inlet (south beach) | 6 | 1972–1993 |
| Vero Beach | 2 | 1979–1984 |
| Fort Pierce (south beach) | 8 | 1971–1995 |
| St. Lucie Inlet | 3 | 1980–1989 |
| Jupiter Island | 14 | 1957–2002 |
| Martin County | 2 | 1996–2002 |

| Beach | Times Nourished | Years |
|---|---|---|
| FLORIDA *continued* | | |
| Jupiter/Carlin Beach | 1 | 1995 |
| Lake Worth Inlet (south beach) | 17 | 1944–1994 |
| Palm Beach | 8 | 1944–1987 |
| Boynton Inlet | 2 | 1961–1973 |
| Delray Beach | 6 | 1973–2002 |
| Boca Raton (north) | 1 | 1988 |
| Boca Raton | 1 | 1998 |
| Boca Raton (south) | 2 | 1985–1996 |
| Pompano Beach/Lauderdale-by-the-Sea | 2 | 1964–1983 |
| John U. Lloyd Beach State Park | 2 | 1977–1989 |
| Hollywood/Hallandale | 2 | 1979–1991 |
| Sunny Isles | 2 | 1988–1990 |
| Haulover Park | 4 | 1960–1987 |
| Bal Harbour | 5 | 1960–1990 |
| Miami Beach | 4 | 1978–1999 |
| Key Biscayne | 2 | 1987–2002 |
| Key West, Smathers Beach | 2 | 1960–2000 |

A replenished beach is indistinguishable from a natural beach in a number of ways. Most of the surface features on natural beaches, ranging from ripple marks and air holes to offshore bars and even barking sand, may be present on artificial beaches, especially after a beach has been shaped by waves for a year or so. But there also are differences. For example, because nourishment involves piling sand on the landward end of the beach profile, this sand is eroded relatively quickly, producing erosional scarps, or beach bluffs. Scarps, signs of rapid erosion, usually develop on natural beaches only after big storms.

The scarps on replenished beaches can exceed 10 feet in height, for example, at West Hampton Dunes, New York, and can be a dangerous obstacle, especially for children and the elderly. This danger was demonstrated in 2002 when then New Jersey Governor McGreevey fell down a scarp while inspecting a replenished beach and broke his leg. From North Carolina south, scarps also can be an obstacle for nesting sea turtles. In general, turtles cannot get over a vertical scarp more than 1 foot high.

Another significant difference between natural and artificial beaches is the types of shells on the beach. The nature of the shell assemblage on a beach depends on the source of the borrowed sediment, but it is often entirely different than that of a natural beach. Artificial beaches composed of sand pumped from a lagoon, estuary, or old lagoonal deposits on the continental shelf will be characterized by abundant oyster shells on mid-Atlantic and southern beaches, and clams and blue mussels on northern beaches. Often lagoonal sand is also very muddy.

Beach aesthetics, such as sand color and sediment size, is also altered with artificial fill. Artificial beaches are usually darker than the original beaches. This color difference is especially true for southern beaches, which are naturally light brown—shells and quartz stained brown—to white—dominated by unstained quartz—in color, while most replenished beaches are light to dark gray. The dark color is from shelf sediment that was buried more than 2 to 3 feet below the seafloor. At that depth, low oxygen content in the sediment causes the grains to be coated with a dark iron compound. A few artificial beaches, like those at Miami Beach, are very light colored because the sand is made up entirely of the calcareous remains of organisms.

Like most of the beaches north of it, Miami Beach originally contained a high percentage of quartz sand. Miami's replenished beach is composed of shell and coral fragments that were dredged from offshore and have a different color, grain size, and composition. The packing together of fragments of shell and coral has resulted in a hard beach surface. Grain size can also alter the amount of turbidity in the nearshore zone and the ability of water

This beach nourishment material at Pine Knoll Shores, on Bogue Banks (North Carolina), was full of large, sharp shell fragments. Quarter for scale.

to percolate through the beach. Even the ambient temperature of the beach can be altered by differences in grain size and color, which is a problem for sea turtle hatchlings. White beaches are more reflective than dark beaches, which absorb more heat; so dark beaches are warmer than lighter-colored beaches under the same environmental conditions.

Some replenished beaches are of very poor quality in terms of sediment size. The most common problem is an overabundance of cobbles or large, broken shell fragments with sharp edges, both of which can make a beach virtually unusable to barefoot swimmers. Such was the case on Pine Knoll Shores (North Carolina), where a 2002 dredge-and-fill replenishment operation broke up the shells, leaving sharp fragments on the beach. The same year at nearby Emerald Isle, the shells were not broken up, but the sediment contained such a high proportion of sharp oyster shells that this beach was also detrimental to walking barefoot. A replenishment at Jacksonville Beach, Florida, produced similar sharp-edged fill material, and a project at Oak Island, North Carolina, left the beach strewn with patches of grapefruit-sized fragments of limestone. In more than one replenishment project beach users were unhappy with such debris, resulting in additional costs to sweep the beaches of coarse rubble.

A beach nourishment project at Oak Island, North Carolina, left the beach strewn with cobble-sized rock fragments. The goal of this project had been to restore turtle habitat, but turtles require sand in which to lay their eggs.

Often a beach with too much coarse shell material can be superficially cleaned by removing the material from the surface of the beach. The problem is that as erosion ensues, the intertidal zone quickly becomes paved with fragmented material that was previously below the beach surface. On such beaches, the zone between high and low tide remains paved with particles that are dangerous to walk on with bare feet.

Much more important than color and shell content is the difference between the live fauna and flora in natural and artificial beaches. Beach critters, from microscopic meiofauna to sand fleas and shell-bearing animals, are all part of the beach ecosystem. The birds feeding in the swash zone depend on these living resources as do many of the fish that are sought by surf fishermen. When borrowed sand is first placed on the beach, these beach-dwelling creatures and life forms are all killed. While visiting the beach during or shortly after nourishment, the odor that you notice, along with flocks of feeding seagulls, is a result of this.

The extent and rate of fauna and flora recovery is not well understood, because the recovery of organisms over long periods on most artificial beaches is not monitored. Based on relatively few studies, it is thought that most organisms reestablish themselves within three years. However, many beaches require re-replenishment at intervals ranging from 3 to 5 years in

Nourishment sediment doesn't always match the specifications of the project plan. This 1998 project at Debidue Beach (South Carolina) placed sediment that was trucked to the beach from a nearby inland borrow pit. The sandy sediment had enough mud and organic material in it to create a wide zone of turbid water in the surf.

order to maintain the beach, and local renourishment of hot spots occurs more frequently. Replenishment intervals of 2 to 3 years or less do not allow organisms enough time to settle in and reestablish the ecosystem. One highly visible sign that a beach has been replenished is a paucity of the birds that dart back and forth, feeding in the swash zone.

Beach replenishment also affects the borrow site where the seafloor is excavated. Dredging destroys offshore plants and animals at the borrow site. Turbidity, which reduces water clarity and fouls the siphons of filter-feeding organisms, occurs both at the borrow site as the sediment slurry leaks from the transport pipe and when the sediment is placed on the beach. The combined effects of altered plant and animal populations, changes in water quality, breaks in the food chain, and related impacts caused by humans and machinery may have a cumulative effect on whether organisms survive at all. Just as we cannot see all the impacts nourishment has on the animals living in the beach, we do not see the impact nourishment has on organisms living offshore.

To fully understand beach replenishment, one needs to understand the political realities behind the process. On the East Coast, replenishment is carried out almost exclusively to save buildings and roads, not to save beaches. As shorelines retreat in response to sea level rise and human interference with sand supply, the beach moves back too. Once buildings are built close to a shoreline, they are threatened by such retreat. Often during the political fracas that precedes a replenishment project, the argument is made that the beach needs to be saved. One shouldn't believe that for a minute. The crisis—perceived of as needing a "solution"—is threatened buildings. The beach does not need saving.

Most artificial beaches are satisfactory for recreational purposes, but many beach visitors, especially beach purists like us, agree that nothing beats a natural beach. Nature knows best!

### *Artificial Dunes*

Dunes are a natural part of many coasts, but like beaches they have suffered the consequences of overdevelopment. In an attempt to restore and maintain the natural environment, artificial dunes are commonly constructed at the shore. One of the most common approaches is to plant dune grass and/or place sand fencing at the back of the beach, both of which trap sand and encourage dune growth. This approach has some merit in that it encourages the natural processes that form dunes and keeps people from tramping through the dunes; but the process is slow, and years of average weather, free of big storms, are required to build high and wide protective dunes.

Large beach nourishment projects usually include efforts to repair or construct dunes through the planting of dune grasses and the use of sand fences.

Building dunes is a good thing for aesthetics and to protect beachfront buildings from storms. Studies, however, have not shown that artificial dunes make much difference in erosion rates. The volume of sand in artificial dunes is very small compared to the total volume of sand a storm moves about on the beach and shoreface.

The most spectacular example of artificial dunes on the East Coast is on the Outer Banks of North Carolina. During the Great Depression of the 1930s, the government created the Civilian Conservation Corps (CCC) as a means to provide both jobs and useful public works. People perceived that the overwashed Outer Banks had an erosion problem, so the CCC constructed a continuous dune line from the Virginia–North Carolina line to Ocracoke. They built the dunes using bulldozers, mules and drag pans, and sand fencing. This dune line sheltered North Carolina Highway 12, which connected the various fishing villages on the islands, from storms. Unfortunately, they built the dune atop miles of broad overwash flats that provided nesting habitat for a wide variety of shorebirds. Fifty miles of beach bird habitat, such as that of the piping plover, was destroyed. The dune line also cut the back sides of the islands off from frequent overwash, so the islands began to erode on the lagoon side as well as the seaward side. After decades of rebuilding Highway 12 at enormous cost, the state of North Carolina is now considering building an elevated causeway along the back side of the Outer Banks.

### *Beach Bulldozing and Sweeping*

Another approach to trying to save beachfront buildings is bulldozing. Bulldozers scrape the beach at the water line, moving sand to the back of the beach in order to build a small sand ridge (artificial dune) in front of buildings. The shallow furrow where the sand was borrowed is supposed to trap additional sand and restore the beach. Bulldozing is usually a sign that a beach is eroding, and only major nourishment will hold it in place. The sand piles produced by scraping are not true dunes, but the faint hope is that they will offer some protection to property during the next storm. Similar to pumping sand onto the beach, scraping also kills the critters living in the sand, as is evidenced by the swooping gulls feeding in the newly piled sand.

Beach grooming, or sweeping with rakes dragged behind tractors, is yet another process that impacts beach biota as well as potentially contributes to sand loss. If our only experience with beaches was at resorts or along heavily developed residential shorelines, our concept of beaches would be that they are sterile. Experienced beachcombers prefer the living beach, one marked by interesting features underfoot. For example, walking a wrack line reveals an array of interesting features, and wrack clearly attracts a variety of beach creatures as a food source and shelter. But you will be hard-pressed to find a wrack line on a manicured resort beach or on many community beaches, which are swept clean daily of any debris that managers deem undesirable. All flotsam and jetsam is removed, including natural constituents such as seaweed, driftwood, and *Spartina* species straw that is flushed out to sea through inlets and washes up onto the beach. These materials provide food and natural habitat for beach creatures. Studies have shown that groomed beaches are barren in comparison to natural beaches, having fewer creatures and less diversity. Removal of wrack also eliminates a beach's natural sand-trapping abilities, further destabilizing the beach. In addition, the sand adhering to seaweed and trapped in the cracks and crevices of wrack is removed from the beach. The amount of sand lost due to a single sweeping is small, but the sum is considerable over a time frame of years. Ironically, we have seen groomed beaches backed by artificial plantings and sand fencing meant to build dunes, so the cost of combating nature is doubled on these beaches.

On many of the beaches of Florida and the urban areas of New York, New Jersey, Delaware, and Maryland, grooming is usually carried out early in the morning or sometimes at night. It's easy to spot the marks of rakes and the perfectly straight plant boundary on the uppermost part of a groomed beach.

Just as many coastal homeowners have a preconceived idea that a coastal lawn should look the same as an inland lawn, some of us have a cultural

Beach cleaning, or sweeping, removes harmful trash but damages natural habitat in a variety of ways; for example, by impacting meiofauna, removing wrack, and destroying nesting grounds. Ironically, sand is also lost in the process, the resource that communities wish to preserve. —Photo by © Sidney Maddock

concept of what a beach or dune field should look like, even if that concept defies nature. It's a struggle to maintain an artificial beach and dune, just as it is a struggle to maintain a classic lawn in the sandy, windswept, salt-air environment at the shore. Which is preferable, a beach of uniform sand raked clean, or a living beach of unexpected variety as reflected by clam and snail shells, crab holes, seaweed, an occasional jellyfish washed up by waves, and pioneer plants popping up along a line of sand-trapping wrack?

## How Beach Resorts Age: The Evolution of Beach Communities

Although beaches constantly renew themselves as they migrate landward with rising sea level, the settlements we build on beaches age relatively quickly. Several studies of the maturation history of beach developments reveal trends in aging that are common for many resorts. This is not to say that all beach resorts go through the same process, but that many have.

The first stage of tourist involvement in what will become a beach resort is exploration. This stage occurred long ago on U.S. East Coast beaches when that first adventurous person visited a remote, rural coastal area and had a wonderful experience. Think of Henry David Thoreau's mid-nineteenth century trip to Cape Cod. There were no tourist accommodations then; lodging was found in a fisherman's or farmer's spare room. No tourist amenities

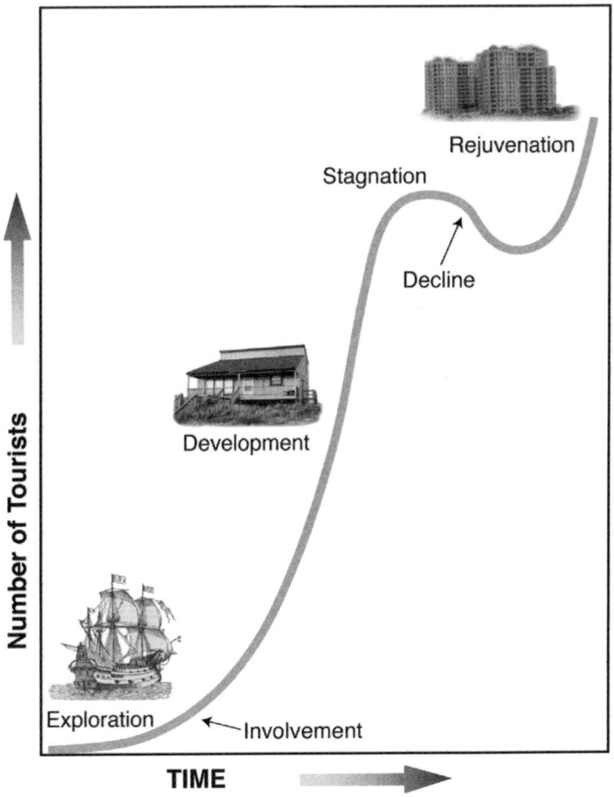

This graph depicts the changing number of visitors over time (years to decades) at a beach resort. —Modified from Meyer-Arendt (1993)

existed, and wildlife and nature provided the only entertainment. For some, this was enough; however, as word spread, more visitors arrived.

New Jersey experienced this exploration stage in the late eighteenth century, as had the Carolinas where plantation beachheads, used by locals in the summer, became the destination of outsiders; for example, people came from Portsmouth, Ocracoke, and Beaufort, North Carolina. Pawleys Island was one of several South Carolina coastal communities with summer homes by the 1790s.

Involvement is the next stage in resort development and occurs as local people begin to take advantage of the increasing number of summer "outsiders." At this stage, local money is invested in seasonal lodging and small-scale recreational activities. In 1801 the first beach advertisement in America appeared in the *Philadelphia Aurora*, touting Cape May, New Jersey, as a beach resort. For the next seventy-five years Cape May was one of the nation's most prestigious beach resorts; it was even the vacation destination of six presidents. Similarly, in 1820, the year Maine became a state, the first bed-and-breakfast opened at Old Orchard Beach; it even included a bowling

This view of Stone Harbor, New Jersey, shows the old dredged ferry route as a long, straight tidal creek that leads to the modern, developed barrier island. The earliest settlements were fishing villages on the lagoons of the area.

alley. Coastal resorts also were developing in Massachusetts and other parts of New Jersey in the 1800s. Pawleys Island had inns and boarding houses by the 1890s, and the involvement stage followed in other communities from Maine to northern Florida. Some of today's well-known resort communities, such as Nags Head and Wrightsville Beach, North Carolina, and Myrtle Beach, Folly Beach, Edisto Island, and Seabrook Island in South Carolina, didn't reach this stage of development until the early 1900s.

At that point in development as a resort, local people had to devise a means to get visitors to the island. Usually, that meant stationing a boat on the mainland to bring people to the island. Eventually, a seasonal ferry service evolved; and finally, bridges and causeways were added. For most barrier islands, the development phase took off with the construction of a bridge or causeway, which improved access and allowed greater numbers of visitors to reach an island.

Access to the beach proved critical for the development stage. In Maine, the railroad provided access to Old Orchard Beach in 1871; by 1886, more than half of the New Jersey coast was accessible by train. To the south this access happened later. For example, rail service didn't connect all the way along the Florida coast to Miami until 1896, and a bridge connecting the mainland to the beach wasn't completed until 1913. Myrtle Beach, South

Aerial photo of Nantasket Beach at Hull, Massachusetts. Every lot is taken in one way or another on this very developed tombolo.

Carolina, wasn't serviced by a modern highway until 1930. The story is virtually the same for every barrier island, many of which didn't reach this phase of the development stage until the 1960s.

Once an efficient means was created to get to the beach, outside money created the means for total beach development in a short span of time. The remaining dunes were bulldozed for more building lots as well as better ocean views. Seawalls and groins were erected following storms to protect properties, and these, in turn, led to further beach erosion. An invasion of fauna and flora from around the world replaced native plants and animals. House cats preyed on defenseless shorebirds. The development phase is complete when just about every lot on a beach is in some way developed. Often sand from bulldozed dunes and mud from dredging operations was used to fill salt marshes behind barrier islands to create more land for buildings. Marshes and wetlands also became sites for the disposal of urban wastes in landfills.

Just as development peaked on many beach resorts, problems caused by development became apparent. The large number of seasonal dwellings coupled with inadequate sewage treatment facilities led to a decline in water and fishing quality. The construction of buildings close to the water's edge eliminated the protective dune and led to property damage. In turn, this resulted in the construction of more engineered structures and further loss of the beach. On some beaches the overall decline in natural attractions led to the construction of amusement parks, but these often attracted what many considered a seedy element not suitable for family beaches.

Thus, the decline in the natural qualities, which had lured people to the beach in the first place, led to a decline in the number of visitors—the decline phase of resort development. Cape May and Atlantic City, New Jersey, reached this point by the 1970s. They were among the first, but others would follow. Decline can continue for a long time (Asbury Park, New Jersey), or there can be a period of rejuvenation.

A number of factors can rejuvenate a beach resort during the rejuvenation phase. A large commitment to beach replenishment, as along the New Jersey coast in the 1990s, can bring investors and home buyers back. This plus casino gambling led to the reconstruction of Atlantic City, although poor neighborhoods remain. A great storm can bring down a large number of deteriorating buildings on the coast and be the impetus for new buildings, which are generally larger than the originals.

Resort communities and their infrastructure usually evolve at the expense of the beach, the raison d'etre for the community. The exceptions are privately owned reserves or undeveloped public lands such as state and national parks. In some states, like Maine, laws have been enacted in an attempt to control unsound and unsafe beach development, but natural beaches remain an endangered species.

The beach at Cape May, New Jersey, was a major tourist resort in the mid-1800s but had gone into decline and even bankruptcy by the late 1960s. Note the shoreline recession south (top of photo) of Cape May, while the community has held its shoreline in place artificially with groins, seawalls, and beach fill. At the time of this photo, the dry beach at high tide was limited to only fillets on the updrift sides of groins.

# Epilogue

We began this book by noting that the most fundamental components of our environment, and of the beach environment in particular, are the geologic and climatic settings. These controls find expression in the regional aspects of the coast and beach types—whether barrier islands occur or rocky coasts—down to the properties of the ripple marks on the surface of the beach. Even the individual grains of sand are controlled by these factors.

We ended the book examining a third environmental component, the human element, which must be considered in any discussion of our environment and beaches. The human attitude toward the natural environment, and our perception of what a beach is or should be, is shaping the future of America's beaches, raising some final questions: Are beaches an endangered species? Are beaches playgrounds or wildlife habitats?

The term "endangered species" conjures up images of sea turtles, piping plovers, and rare plants that many of us have never seen or valued. The term implies a sense of urgency, a need for action. Beaches are not living organisms, yet we characterize them with the terminology of life. We discuss their life cycle using terms such as equilibrium, a healthy beach, a nourished beach, or a groomed beach. For example, beach equilibrium implies a balance between sand supply, sea level, wave conditions, and the shape of the beach. Change any one of these and the beach, reflexively, changes. By analogy, beaches are a unique natural "species" that has attracted us to the sea's edge.

The second question above relates to the core of our perception of beaches. From our first experiences frolicking on a summer strand, we have enjoyed beaches as playgrounds. With maturity, the nature of the "play" may have changed from pails and shovels to binoculars and metal detectors, but how many of us really think of a beach as habitat? That question looms larger as competition for beach use, whether it is replenishment versus turtles or grooming versus birds, increases. If beaches become endangered as natural habitats, then many elements of our play on the beach, from fishing to even walking, may become endangered as well. It is already happening.

## The Future of Beaches: Are Beaches an Endangered Species?

Americans are rushing to the Atlantic Ocean to fulfill their dreams of living as close to a shoreline as possible. Judging from magazine and TV advertisements, the shorefront beach cottage is the ultimate evidence of success for some, the perfect place to park a sports car and drink a café au lait in the morning and Caribbean rum in the evening. Meanwhile, the beach is cooperating to fulfill this dream of closeness as it inexorably retreats toward the houses.

The beaches of the world are changing, and U.S. East Coast beaches are probably changing as fast as any. More and more beaches are artificial; they have been nourished and replenished into a state unlike their original one. Fewer and fewer beaches offer an uninterrupted vista of the surf zone without the visual intrusion of high-rise buildings, oversized "cottages," piers, power lines, and other visual infrastructure. Even the landward edges of more and more beaches are lined with steel, wood, or stone seawalls and groins, large and small, that run across the beach and into the surf zone. Worst of all, many beaches have narrowed up against seawalls to the point that tide-tables are required to safely go swimming; otherwise, there might not be a beach. Because the sun sets in the west, it is increasingly difficult to sunbathe in the afternoon on some East Coast beaches because of the shadows of high-rise buildings!

We first observed "tide-table swimming" at Cape May, New Jersey, a few decades back. Since no beach existed in front of the seawall at mid- to low tide, it was necessary to be poised for swimming at the only time the beach was present—the four hours around low tide. For the rest of the Cape May tidal cycle, the principal entertainment, outside of local bars, was promenading on top of the huge seawall.

A piping plover (*Charadrius melodus*) chick depends on the wrack line for cover and food. Atlantic beaches provide critical habitat for this species as well as many other plants and animals. The decline in piping plovers along the Atlantic Coast is related to increased development. —Photo by © Sidney Maddock

We argue that beaches should be preserved for all sorts of reasons. One of those reasons has to do with the fact that the beach is a beautiful but complicated ecosystem. One of the biggests threats to the beach ecosystem is beach nourishment. The microscopic meiofauna between sand grains, sand fleas, mole crabs, gulls, plovers, terns, shellfish, and bottom fish that live near the surf zone are members of an interconnected community that depend on one another for food. When the system is interrupted and suffocated by a nourished beach, the people who fish in the surf zone will be the first to detect a change because the fish have moved on to greener pastures—ones with more food.

One of the most important reasons for beach preservation, as far as people are concerned, is the beach's recreational value. The fact that you are reading this book means that there is no need to justify this fact to you. From a recreational standpoint, a threat to the health of beaches is shorefront construction. The cottages and high-rises themselves may not damage the beach, but defending them from erosion often completely destroys the beach. Construction of seawalls is the worst example, but anything built next to a beach, be it a road or a row of closely spaced buildings, causes it to narrow and eventually disappear. The truth is, seawalls destroy the beaches used by a large number of people to save the property of a very small number of people. Buildings are defended at all costs, including the loss of the ecosystem, beach quality, and aesthetics.

Unfortunately, there is little room for political compromise at the shoreline. If we are to prevent beach loss, we can only move buildings back or demolish them. Efforts to hold the beach in place will almost always lead to its degradation or destruction in the long run. A long history of beach management tells us that once a seawall is put in, it is never taken out. The wall just grows bigger with time. Beach nourishment is a perpetual process just like upgrading seawalls. Once it starts, renourishment goes on and on, every three to eight years or so.

What will East Coast beaches be like one hundred years from now? If we assume current management policies prevail, we can expect to find more of the East Coast shoreline lined with high-rise condos and hotels. Most beach communities will be densely populated (in summer) with strong commercial bents, like today's Atlantic City, New Jersey; Ocean City, Maryland; Myrtle Beach, South Carolina; and Miami Beach, Florida. With the exception of the few stretches of national seashore and coastal state parks, the quiet, contemplative, remote, and peaceful aspects will be gone. Most beaches will be nourished, but funding for nourishment will come solely from local tax bases, because federal and state taxpayers will eventually recognize that they should not have to pay to save buildings placed next to eroding shorelines. As sea level rises, erosion rates will increase, particularly on nourished beaches, and the cost of nourishment will increase significantly. Very likely, many communities will be forced to maintain, by nourishment, a short, centrally located beach for swimmers while the rest of the shoreline will have large seawalls with waves breaking on them even at low tide.

It's not a pretty picture to contemplate, and we hope that people will educate themselves and make their voices heard in defense of the beach!

### *Fifteen Ways to Conserve a Beach for Future Generations*

The following suggestions are a beginning to maintaining our Atlantic beaches underfoot.

1. Recognize that beaches are a very limited recreational resource that belong to all people.
2. Recognize that beaches are more important than buildings.
3. In all actions regarding beach management, remember that sea level is rising and the rise is likely to accelerate, as will shoreline retreat.
4. In all actions regarding beach management, consider the long view (five generations).
5. Site new buildings as far away from the beach as possible.
6. If buildings are threatened, move them back, demolish them, or let them fall into the sea.
7. Don't allow high-rise buildings near a beach.
8. Don't nourish beaches.
9. Don't allow seawalls or other shore-hardening structures to be built.
10. Discourage coastal engineering on adjacent beaches.
11. Don't drive on beaches.
12. Don't rake and sweep beaches to groom them.
13. Do educate young and old about the value of beaches.
14. Do encourage governmental agencies and conservation-oriented nonprofits to purchase the remaining wild and undeveloped beaches to preserve them.
15. Do maintain beach-community monitoring programs; for example, tracking changes in the dynamics of the beach ecosystem such as beach width, sand supply, biodiversity, natural productivity, and water quality.

A piping plover, an endangered species, finds its lunch in the beach. Are Atlantic Coast beaches also an endangered species? —Photo by © Sidney Maddock

# Glossary

**accretion.** The accumulation of sediment on a beach that causes it to widen and build seaward. The opposite of erosion.

**adhesion structure.** A bed form that is produced when dry sand is blown over a damp surface. The sand sticks to and accumulates on the surface to form irregular, wartlike features, or small, irregular ripples.

**antidune.** A long, low, asymmetrical ridge of sand, several inches high, with a steep face on the upstream side of the dune that faces the direction of flow. They form on beaches as backwash flows seaward over fine-grained sand, usually on a steep portion of the beach, but are washed out by the falling tide.

**armored mudball.** See **mud ball**.

**barrier island.** A long, narrow, sandy island that faces the open ocean, parallels the mainland, and is bounded on either end by inlets. Most of the East Coast, from the South Shore of Long Island to Miami Beach, consists of barrier islands.

**barrier island migration**. The landward movement of an entire barrier island in response to rising sea level or a loss of sand supply.

**beach cusps.** See **cusps**.

**beach grooming.** Raking or sweeping of the beach to remove wrack and debris.

**beach nourishment.** The method of holding a shoreline in place by adding sand from an outside source. Also called *beach replenishment* or *dredge-and-fill*.

**beach scraping.** A procedure in which a bulldozer pushes a layer of sand from the intertidal zone to the back of the beach to form an artificial dune.

**bed form.** A small-scale feature that forms on the surface of a beach or dune. Bed forms may be buried and preserved in a beach's structure or wiped out by the next wave, tide, or gust of wind. Also called *sedimentary structures*. Examples include lineations, rills, and ripples.

**berm.** A depositional, terracelike feature that forms on the upper beach.

**berm crest.** The top of a berm that separates the steep, seaward-facing berm face from the gently sloping, landward side of the berm.

**black sand.** See **heavy minerals**.

**blowout.** A flat or bowl-shaped area in a dune or dune field where sand has been blown away. The wind often erodes down to the level of the water table where wet sand prevents further erosion.

**breakwater.** See **offshore breakwater**.

**calcareous.** Any material composed of calcium carbonate. On beaches, calcareous materials are usually shells and skeletal material, such as coral, and the sand derived from these sources.

**Coastal Plain.** The physiographic province of the East Coast that extends from Long Island to South Florida and is underlain by sedimentary rocks that dip gently seaward. Its shoreline is characterized by extensive barrier island chains.

**continental shelf.** The gently sloping surface of the edge of the continent that extends from the beach to where the steeper continental slope begins, usually at ocean depths greater than 300 feet.

**coquina.** Shell material, usually including the common coquina clam (*Donax* species), that has been cemented with calcium carbonate to form rock.

**crescent mark.** The scour mark around an object on the beach, such as a shell or pebble, resulting from wind or water flow around the object.

**cross-bedding.** The inclined beds that form in a dune on its sloping faces and dip downward in the direction of sediment transport.

**current ripple.** A type of asymmetrical ripple mark that forms as water flows on the beach, for example, in a trough. The steep face of the ripple faces downstream.

**cusps.** Regularly spaced embayments of variable size separated by small ridges, or horns, along the beach. They give the shoreline a sinuous appearance.

**down-drift.** The direction of longshore current flow; it is analogous to *downstream* in rivers. It is the dominant direction of sand movement on beaches.

**dredge-and-fill.** See **beach nourishment**.

**dredging.** The removal of sand by dredges to improve a navigational channel for boat traffic or to provide sand for beach nourishment.

**dry beach.** The portion of a beach that is dry at high tide.

**dune.** A hill or ridgelike feature, either bare or covered with vegetation, that is an accumulation of windblown sand.

**ebb current.** The tidal current that forms when the tide is going out.

**ebb tidal delta.** A body of sand that protrudes seaward of an inlet and was formed by ebb tidal currents.

**erosion.** In terms of the shoreline, it is the net loss of sand from a beach that leads to the retreat of the shoreline.

**estuary.** A river valley that has been flooded by a rising sea and is a place where fresh- and saltwater mix.

**fetch.** The distance of open water over which the wind can blow to form waves.

**flood current.** The tidal current that forms as the tide is rising.

**flood tidal delta.** A body of sand that protrudes landward of an inlet and was formed by flood-tidal currents.

**foredune.** The dune closest to the beach.

**foreshore.** The seaward-dipping zone on a beach from the crest of the seaward-most berm to the low tide levels.

**groin.** An engineered structure installed perpendicular to the beach in an effort to trap sand traveling with the longshore current. Shorter than jetties, groins are almost always placed in groups, or fields. Groins cause sand accretion on the updrift side but erosion on the down-drift side.

**grooming.** See **beach grooming**.

**groundwater.** Underground water in the zone of saturation. On a beach, the groundwater table meets the sea to form springs and seeps and influences coastal processes.

**heavy minerals.** The mineral fraction of beaches and dunes that consists of grains that are denser than quartz and feldspar (the light minerals). Heavy-mineral concentrations are deposited during storms or are formed by the winnowing effect of wind and water. These minerals are often darker than the light minerals and impart different colors to the beach or dune, for example, black sand from magnetite and reddish brown sand from garnet. Also called *black sand*.

**inner bar.** The landward sandbar of two or more submerged sandbars located off a beach.

**inlet.** The narrow waterway between two barrier islands that connects the sea and a lagoon.

**interference ripple.** A complex ripple mark produced by waves or currents that flow in more than one direction.

**intertidal zone.** The wet portion of the beach exposed at low tide. The zone between low and high tides.

**jetty.** A long structure built perpendicular to the shoreline that is usually placed on both sides of inlets and river mouths to stabilize a navigational channel and prevent sediment in longshore currents from clogging the channel.

**ladder-back ripple.** A complex ripple mark that is produced when one set of ripple marks forms atop a previous set at nearly a right angle, producing a ladder pattern. Changing wave directions during the falling tide produce them, and they are usually found on tidal flats.

**lagoon.** The body of water between a barrier island and the mainland.

**longshore current.** The current that flows parallel to a beach; waves striking the beach at an angle generate this current.

**macrofauna.** Animals that are visible to the naked eye.

**médano.** A Spanish term for a large solitary dune, especially one along a seashore; for example, Jockey's Ridge (North Carolina).

**meiofauna.** Microscopic animals that live between sand grains in the beach.

**migration.** See **barrier island migration**.

**mineral suite.** As applied to sand, it is a unique assemblage of minerals that characterizes beach, river, or glacial outwash sand and reflects the composition of its source rocks as well as history of weathering and transport.

**mud ball.** A ball-shaped clump of clay of variable size that forms when clay outcrops are eroded, or when mud is pumped onto the beach during nourishment projects. A mud ball that has rolled over coarser material to become coated with sand, shell fragments, or gravel that slows the ball's attrition is called an *armored mud ball*.

**nail hole.** An informal term for holes in the beach that are produced by escaping air or the truncation of air cavities in the beach. They are so named because their size range is similar to that of nails. Also called *sand holes*.

**neap tide.** The minimum tidal range at a beach that occurs during the first and third quarters of the moon.

**nourished beach.** An artificial beach that has been widened or maintained by the importing of new sediment.

**nourishment.** See **beach nourishment**.

**overwash.** Beach sand that storm waves have transported inland beyond the beach.

**overwash fan.** An accumulation of sand and shells that is deposited by overwash, usually in a fan shape.

**offshore bar.** An underwater ridge of sand off the beach that is often identifiable by a line of breaking waves.

**offshore breakwater.** An engineered structure that is placed offshore and parallel to the beach. Breakwaters mimic sandbars and cause waves to break, lessening erosion on the beach behind them. But they interrupt the longshore current and cause erosion on down-drift beaches.

**outer bar.** The outermost sandbar of two or more sandbars on a beach. The biggest waves break on this seaward-most bar.

**pedestal structure.** A toadstool-shaped feature that forms as a result of the differential wind erosion of damp and dry sand layers.

**Piedmont.** The regional plateau lying east of the Appalachian Mountains and west of the Coastal Plain from New Jersey to Georgia. Much of the sand on U.S. Atlantic Coast beaches, from New Jersey to northern Florida, was ultimately derived from the Piedmont region of igneous and metamorphic rocks through river transport.

**Pleistocene.** A unit (epoch) of geologic time that spans from about 2 million to 10,000 years ago; also called the Ice Age.

**plunging breaker.** A wave that breaks on a moderate beach slope (usually 3 to 11 degrees). A plunging breaker curls over, forming a barrel or tube of air as it collapses. It's the most forceful type of breaker in terms of generating sand movement on the seafloor.

**provenance.** A place of origin; the area from which sand grains in rivers, beaches, and dunes were derived.

**replenishment.** See **beach nourishment**.

**rhomboid ripple mark.** A diamond-shaped ripple mark that wave swash creates during falling tides.

**rill mark.** A small erosional channel in the sand carved by either fresh- or saltwater draining out of the beach at low tide. At the end of each rill the sand is deposited in small deltalike features.

**ring structure.** A ring pattern on the beach surface that is produced when water truncates a blister or arched sand layer.

**rip current.** A narrow, fast-moving water current flowing seaward from the beach through the surf zone. It is a major hazard for swimmers.

**ripple mark.** A small ridge and depression in the sand; they typically occur in a repetitive pattern. Different patterns are created by different air, water, and wave conditions. In cross section, if both faces of a ripple have approximately the same slope, the ripple is said to be *symmetrical*. Waves typically produce symmetrical ripples. Most ripples have a steeper face in the downstream direction and are asymmetrical in cross section. Wind and water currents and some waves produce asymmetrical ripples.

**runnel.** See **trough**.

**salt marsh.** A sandy mudflat covered by salt grasses. It occurs in protected, quiet waters.

**sand.** Grains of material, such as minerals like quartz or magnetite or bits of shells, that range in size from 1⁄16 to 2 millimeters in diameter.

**sandbar.** See **inner bar**, **offshore bar**, and **outer bar**.

**sand hole.** See **nail hole**.

**scarp.** A sand bluff on the beach that parallels the shore and usually indicates rapid erosion. They are very common on nourished beaches.

**sea.** The choppy, irregular (confused) water surface that forms when waves are generated locally.

**sea cliff.** A cliff or steep slope cut in bedrock at the back of a beach or at the sea's edge. In contrast, a bluff is eroded in unconsolidated sediment.

**Sea Island.** A term for barrier islands of the Georgia and southern South Carolina coasts that consist of Pleistocene and modern islands that have been joined together.

**sea level.** The average elevation of the surface of the sea.

**sea stack.** A nearly vertical rock mass that formed when it became detached from the mainland as the result of wave erosion. They are found along rocky shores.

**seawall.** An engineered wall installed on the upper beach parallel to the beach in an effort to prevent the retreat of the shoreline and erosion of property.

**sedimentary structure.** See **bed form**.

**shell hash.** A concentration of broken shell material on a beach.

**shoreface.** The narrow, relatively steep surface that extends seaward from the beach, often to a depth of 30 to 60 feet, at which point the slope flattens and merges with the continental shelf. The shoreface is the surface of active sand exchange between the inner continental shelf and the beach.

**spilling breaker.** A wave that breaks on a relatively flat beach (typically 3 degrees or less). The wave crest literally spills over the top of the wave but does not curl like a plunging breaker.

**spit.** A fingerlike extension of the beach that was formed by longshore sediment transport; typically, it is a curved or hooklike sandbar extending into an inlet.

**spring tide.** The highest tidal range at a beach that occurs during a full or new moon.

**storm surge.** The rise in water level due to low atmospheric pressure at the center of a storm; the water mounding due to circulation around the low-pressure center; and the effect of water being pushed into shallower depths and onshore.

**surf zone.** The area between the outer line of breaking waves and the landward limit of wave uprush.

**surging breaker.** A wave that comes ashore on a steep beach (generally greater than 11 degrees) and breaks directly on the beach.

**swash.** A thin layer of water that is the final remains of a wave as it rolls up the beach.

**swash mark.** A sinuous line of sand or shell debris that wave swash left at its uppermost limit.

**swash zone.** The area of the beach over which wave swash is running up and down the beach.

**swell.** Evenly spaced waves with long wavelengths that are formed by winds far from the beach.

**tide.** The daily elevation and depression of local water level caused by the gravitational pull of the sun and moon on the ocean's waters as the earth rotates.

**tidal range.** The vertical difference between normal high and low tides. Also called *tidal amplitude*.

**tombolo.** A sand or gravel bar that connects an island to the mainland or to another island.

**trough.** A feature that forms between berm crests. A trough is usually visible as a drainage channel at low tide and is a place where current ripples form.

**updrift.** The direction that is the opposite of the longshore current; it is analogous to *upstream* in rivers and is the opposite of *down-drift*.

**water table.** The surface of the zone of groundwater saturation.

**wave.** The form water takes as energy is transferred from the wind to the sea surface. It consists of a crest (high point) and trough (low point) and moves through the water from its wind source to a coastline. Waves move water in a circular or elliptical rotation.

**wave amplitude.** Half of the vertical distance between the wave crest and trough.

**wave height.** The vertical distance between the wave crest and wave trough.

**wavelength.** The distance between wave crests.

**wave orbital.** The internal water movement caused by the passage of a wave. Orbitals are circular in deep water and elliptical in shallow water.

**wave period.** The time it takes for a wave crest to pass a given point.

**wave refraction.** The bending of waves as they come ashore, begin to feel bottom, and slow down.

**wave ripple.** A type of ripple mark formed by waves and also referred to as a *long-crested ripple mark*.

**wave trough.** The lowest point of a wave.

**wet-dry line.** The point on the beach where the intertidal zone (wet) part of the beach ends and the dry beach that does not get inundated at high tide starts.

**wind ripple mark.** A ripple mark that is formed by wind, has long crests, and usually is shorter than ripples formed by water currents and waves. Typically, they form on the back of the beach and in sand dunes.

**wrack line.** A line of natural and artificial debris—for example, seaweed, *Spartina* straw, fishing nets, lumber, driftwood, and plastic bottles—marking the previous landward extent of the high tide and/or wave swash.

# References

Allen, J. R. L. 1982. *Sedimentary Structures: Their Character and Physical Basis*. Amsterdam, Netherlands: Elsevier.

Bertness, M. D. 1999. *The Ecology of Atlantic Shorelines*. Sunderland, MA: Sinauer Associates, Inc.

Bird, E. C. F. 1993. *Submerging Coasts: The Effects of a Rising Sea Level on Coastal Environments*. Chichester, England: John Wiley & Sons.

Bush, D. M., Longo, N. J., Neal, W. J., Esteves, L. S., Pilkey, O. H., Pilkey, D. F., and C. A. Webb. 2001. *Living on the Edge of the Gulf: The West Florida and Alabama Coast*. Durham, NC: Duke University Press.

Bush, D. M., Neal, W. J., Longo, N. J., Lindeman, K. C., Pilkey, D. F., Esteves, L. S., Congleton, J. D., and O. H. Pilkey. 2004. *Living with Florida's Atlantic Beaches: Coastal Hazards from Amelia Island to Key West*. Durham, NC: Duke University Press.

Carter, R. W. G. 1988. *Coastal Environments*. London, England: Academic Press.

Clayton, T. D., Taylor, L. A., Jr., Cleary, W. J., Hosier, P., Graber, P. H. F., Neal, W. J., and O. H. Pilkey, Sr. 1992. *Living with the Georgia Shore*. Durham, NC: Duke University Press.

Davis, R. A., Jr., ed. 1994. *Geology of Holocene Barrier Island Systems*. New York, NY: Springer-Verlag.

Davis, R. A., Jr., and D. M. FitzGerald. 2004. *Beaches and Coasts*. Malden, MA: Blackwell Publishing.

Fox, W. T. 1983. *At the Sea's Edge: An Introduction to Coastal Oceanography for the Amateur Naturalist*. NY: Prentice Hall.

Kaufman, W., and O. H. Pilkey. 1983. *The Beaches are Moving: The Drowning of America's Shoreline*. Durham, NC: Duke University Press.

Kelley, J. T., Kelley, A. R., and O. H. Pilkey, Sr. 1989. *Living with the Maine Coast*. Durham, NC: Duke University Press.

Komar, P. D. 1998. *Beach Processes and Sedimentation*. 2nd ed. Upper Saddle River, NJ: Prentice Hall.

Lencek, L., and G. Bosker. 1998. *The Beach: The History of Paradise on Earth*. New York, NY: Penguin.

Lennon, G., Neal, W. J., Bush, D. M., Pilkey, O. H., Stutz, M., and J. Bullock. 1996. *Living with the South Carolina Coast*. Durham, NC: Duke University Press.

McCormick, L., Pilkey, O. H., Jr., Neal, W. J., and Pilkey, O. H., Sr. 1984. *Living with Long Island's South Shore*. Durham, NC: Duke University Press.

Meyer-Arendt, K. J. 1993. Geomorphic Impacts of Resort Evolution along the Gulf of Mexico Coast: Applicability of Resort Cycle Models. In *Tourism vs. Environment: The Case for Coastal Areas*, ed. P. P. Wong, Geojournal Library Series vol. 26. Dordrecht, Netherlands: Kluwer Academic Publishers.

NOAA, National Ocean Service, Coast and Geodetic Survey. 1995. *U.S. East Coast, New Hampshire-Massachusetts-Maine, Portsmouth to Cape Ann*, number 13278, 1:80,000.

Nordstrom, K. F. 2000. Beaches and Dunes of Developed Coasts. New York, NY: Cambridge University Press.

Nordstrom, K. F., Gares, P. A., Psuty, N. P., Pilkey, O. H., Jr., Neal, W. J., and O. H. Pilkey, Sr. 1986. *Living with the New Jersey Shore*. Durham, NC: Duke University Press.

Oldale, R. N. 1992. *Cape Cod and the Islands: The Geologic Story*. East Orleans, MA: Parnassus Imprints.

Pilkey, O. H. 2003. *A Celebration of the World's Barrier Islands*. New York, NY: Columbia University Press.

Pilkey, O. H., Bush, D. M., and W. J. Neal. 2000. Lessons from Lighthouses: Shifting Sands, Coastal Management Strategies, and the Cape Hatteras Lighthouse Controversy. In *The Earth Around Us: Maintaining a Livable Planet*, ed. J. S. Schneiderman. New York, NY: W. H. Freeman & Company. (Republished in 2003 by Westview Press, Boulder, CO.)

Pilkey, O. H, and K. L. Dixon. 1996. *The Corps and the Shore*. Washington, DC: Island Press.

Pilkey, O. H., Neal, W. J., Riggs, S. R., Webb, C. A., Bush, D. M., Pilkey, D. F., Bullock, J., and B. A. Cowan. 1998. *The North Carolina Shore and Its Barrier Islands: Restless Ribbons of Sand*. Durham, NC: Duke University Press.

Pilkey, O. H., Rice, T. M., and W. J. Neal. 2004. *How to Read a North Carolina Beach: Bubble Holes, Barking Sands, and Rippled Runnels*. Chapel Hill, NC: University of North Carolina Press.

Puleo, J. A., and D. Huntely, eds. 2006. Swash-Zone Processes. *Continental Shelf Research* 26 (no. 5, special issue).

Reineck, H. E., and I. B. Singh. 1975. *Depositional Sedimentary Environments*. New York, NY: Springer-Verlag.

Schwartz, M. L., ed. 2005. *Encyclopedia of Coastal Science*. Dordrecht, Netherlands: Springer.

Strahler, A. N. 1988. *A Geologist's View of Cape Cod*. East Orleans, MA: Parnassus Press.

Williams, S. J., Dodd, K., and K. K. Gohn. 1991. Coasts in Crisis. *U.S. Geological Survey Circular* 1075.

Woodroffe, C. D. 2002. *Coasts: Form, Process and Evolution*. Cambridge, England: Cambridge University Press.

# Index

**WILLIAM J. NEAL** is professor emeritus and past chairman of the Department of Geology at Grand Valley State University in Michigan. As a sedimentologist he has been involved in coastal studies since the 1970s. In 1993 he received (with Orrin H. Pilkey) the American Geological Institute's Award for Outstanding Contribution to Public Understanding of Geology.

**ORRIN H. PILKEY** is the James B. Duke professor emeritus in the Nicholas School of the Environment and Earth Sciences at Duke University. He began his career as a deep-sea oceanographer and ended up quite happily on the beach. He received the Francis Shepard Medal for excellence in marine geology and was the 2003 recipient of the Priestly Award, which honors an outstanding American scientist for discoveries that contribute to the welfare of humankind.

**JOSEPH T. KELLEY** is a native of the Maine coast and has a PhD in Geology from Lehigh University. He taught briefly at the University of New Orleans and then became the state marine geologist with the Maine Geological Survey in 1982. In 1999 Joe became a professor of marine geology at the University of Maine. He is currently chairman of the Department of Earth Science there.